THE BEST TEST PREPARATION FOR THE

SAT II: Subject Test

BIOLOGY

Jay Templin, Ed.D.
Assistant Professor of Biology
Widener University, Wilmington, Delaware

Judith A. Stone, Ph.D.
Adjunct Assistant Professor of Biology
Suffolk County Community College, Selden, New York

Clarence C. Wolfe, Ph.D.
Professor & Assistant Division Chairperson of Biology
North Virginia Community College, Annandale, Virginia

Biology Course Review by
William Uhland, Research Scientist

Research & Education Association
61 Ethel Road West • Piscataway, New Jersey 08854

The Best Test Preparation for the
SAT II: SUBJECT TEST IN BIOLOGY

Printed in the United States of America

Library of Congress Catalog Card Number 98-65425

International Standard Book Number 0-87891-644-X

Research & Education Association
61 Ethel Road West
Piscataway, New Jersey 08854

REA supports the effort to conserve and protect environmental resources by printing on recycled papers.

CONTENTS

THE PRACTICE TESTS

PREFACE

This book provides an accurate and complete representation of the SAT II: Subject Test in Biology. The six practice exams and the course review section are based on the format of the most recently administered tests. Each practice test is one hour in length and includes every type of question that can be expected on the actual exam. Following each exam is an answer key complete with detailed explanations designed to clarify the material for the student. By studying the review section, completing all six tests, and studying the answers that follow, students can discover their strengths and weaknesses, and thereby become well prepared for the actual exam.

ABOUT THE TEST

The SAT II: Subject Test in Biology is offered three times during the academic year, in October, January, and May. Students can obtain registration forms and test date information from their school's guidance or counseling office, or they can contact:

College Board SAT Program
P.O. Box 6200
Princeton, NJ 08541-6200
609-771-7600
sat@ets.org

or they can visit the College Board website at:

http://www.collegeboard.org

Although some colleges require these tests as part of the admission process, most colleges will use the SAT II: Biology score for student placement within the biology program.

The test is one hour in length and consists of 95 to 100 multiple-choice questions designed to measure your knowledge of biology and your ability to apply that knowledge. Examinees are expected to have taken at least one year of college-preparatory biology, preferably involving laboratory experience, to be prepared for the types of questions that appear on the examination. It is also advisable for examinees to have a basic understanding of algebra, such as direct and indirect functions and ratios, for computing word problems (the use of calculators is not permitted during the exam).

While emphasis may change slightly from administration to administration, the topics covered on the test include:

Cellular and Molecular Biology	30%
Organismal Biology	30%
Ecology	15%
Evolution and Diversity	15%
Classical Genetics	10%

Since all high school curriculums are not identical, it is not expected that examinees will be familiar with every topic presented on the test.

The exam is designed to test knowledge of biology in three ways:

Concept Knowledge (35%) - Examinees are tested for their ability to remember specific information, facts, and definitions.

Application (35%) - Examinees are required to rework information into proportional formulations, work problems that may need to be solved mathematically, and use information in situations which may be uncommon or applicable.

Interpretation (30%) - Examinees are expected to draw inferences or deductions from data (qualitative and quantitative), identify assumptions that are unstated, and combine information to form a conclusion.

ABOUT THE REVIEW SECTION

This book contains review material that students will find useful as a study aid while preparing for the test. By reviewing each item, students will be able to become better acquainted with information that will most likely appear on the actual test. Included in this section are the following topics:

Chapter 1 - Chemistry

General Chemistry
Biochemical Pathways
Molecular Genetics

Chapter 2 - The Cell

Prokaryotic Cells
Eukaryotic Cells
Cellular Division

Principles of Ecology
Heredity

Chapter 3 - Organismal and Population Biology
The Interrelationship of Living Things
Diversity and Characteristics of the Kingdoms Monera
The Kingdom Plantae
The Kingdom Animalia
Evolution

SCORING THE TEST

The exam is scored by crediting each correct answer with one point and deducting only partial credit (one-fourth of a point) for each incorrect answer. There is neither a credit nor a deduction for questions that are omitted. The resulting value is a "raw score." Use the formula below to calculate a raw score:

_____ — (_____ X 1/4) = _____
number right — number wrong* — raw score (round to nearest whole #)

* DO NOT INCLUDE UNANSWERED QUESTIONS

You may convert your raw score to a scaled score by using the conversion chart on the following page. Your scaled score will be used by your college of choice to determine placement in their biology program. This score will vary from college to college, and is used with other academic information to determine placement. Contact your college admissions office for more information regarding its use of scores. The profile of examinees taking the Biology test in 1996 is as follows:

POINTS SCORED	NUMBER OF STUDENTS	PERCENTAGE
700-800	9,250	17
600-699	18,971	35
500-599	15,435	30
400-499	7,679	14
300-399	2,649	3
200-299	30	1

TOTAL NUMBER STUDENTS TAKING EXAM 52,909
MEAN POINT SCORE 596
STANDARD DEVIATION 100

SCORE CONVERSION CHART

RAW SCORE	SCALED SCORE	RAW SCORE	SCALED SCORE
100	800	45	500
99	790	44	490
98	790	43	490
97	780	42	480
96	780	41	480
95	770	40	470
94	760	39	470
93	760	38	460
92	750	37	450
91	750	36	450
90	740	35	440
89	740	34	440
88	730	33	430
87	730	32	430
86	720	31	420
85	720	30	420
84	710	29	410
83	700	28	410
82	700	27	400
81	690	26	390
80	690	25	390
79	680	24	380
78	680	23	380
77	670	22	370
76	670	21	370
75	660	20	360
74	660	19	360
73	650	18	350
72	640	17	350
71	640	16	340
70	630	15	330
69	630	14	330
68	620	13	320
67	620	12	320
66	610	11	310
65	610	10	310
64	600	09	300
63	600	08	300
62	590	07	290
61	590	06	290
60	580	05	280
59	570	04	270
58	570	03	270
57	560	02	260
56	560	01	260
55	550	00	250
54	550	-01	250
53	540	-02	240
52	540	-03	240
51	530	-04	230
50	530	-05	230
49	520	-06	220
48	510	-07	210
47	510	-08	210
46	500	-09 through -25	200

BIOLOGY

Course Review

CHAPTER 1

CHEMISTRY

A. GENERAL CHEMISTRY

The Elements

Element – An element is a substance which cannot be decomposed into simpler or less complex substances by ordinary chemical means.

Compound – A compound is a combination of elements present in definite proportions by weight. These are substances which can be decomposed by chemical means.

Mixtures – Mixtures contain two or more substances, each of which retains its original properties and can be separated from the others by relatively simple means. They do not have a definite composition.

Atoms – Each element is made up of one kind of atom. An atom is the smallest part of an element which can combine with other elements. Each atom consists of:

A) Atomic Nucleus – Small, dense center of an atom.

B) Proton – Positively charged particle of the nucleus.

C) Neutron – Electrically neutral particle of the nucleus.

D) Electron – Negatively charged particle which orbits the nucleus. In normal, neutral atoms, the number of electrons is equal to the number of protons.

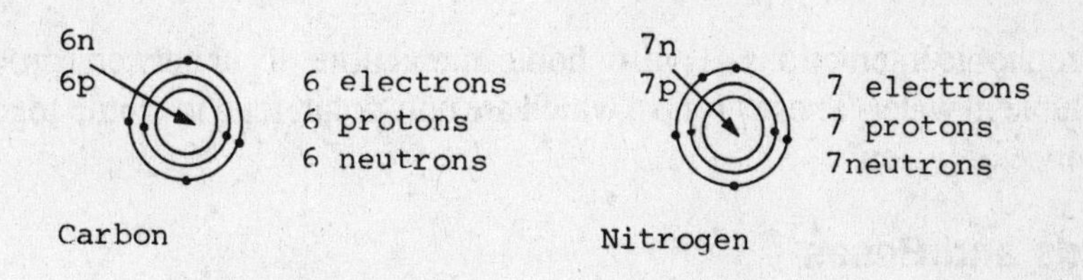

Atomic structure of carbon and nitrogen.

Ions – Atoms or groups of atoms which have lost or gained electrons are called ions. One of the ions formed is always electropositive and the other electronegative.

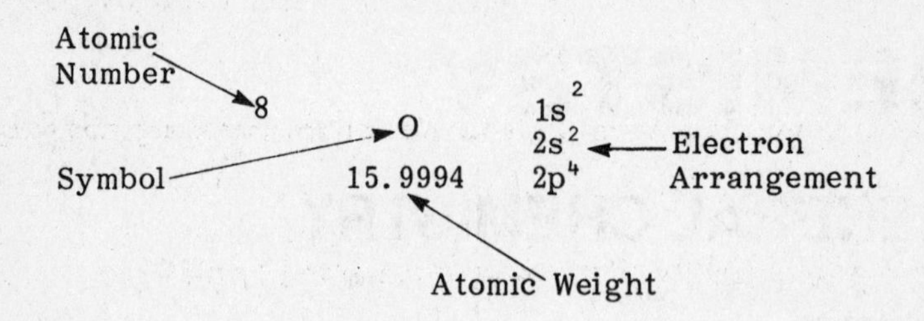

Representation of oxygen from Periodic Table of the Elements.

Chemical Bonds

Ionic Bond – The complete transfer of an electron from one atom to another. Ionic bonds form between strong electron donors and strong electron acceptors.

Covalent Bond – The sharing of pairs of electrons between atoms. Covalent bonds may be single, double, or triple.

Polar Covalent Bond – A polar covalent bond is a bond in which the charge is distributed asymmetrically within the bond.

Non-Polar Covalent Bond – A non-polar covalent bond is a bond where the electrons are pulled exactly equally by two atoms.

Hydrogen Bond – A hydrogen bond is formed when a single hydrogen atom is shared between two electronegative atoms, usually nitrogen or oxygen.

Van der Waals Forces – Van der Waals forces are weak linkages which occur between electrically neutral molecules or parts of molecules which are very close to each other.

Hydrophobic Interactions – Hydrophobic interactions occur between groups that are insoluble in water. These groups, which are non-polar, tend to clump together in the presence of water.

Acids and Bases

Acid – An acid is a compound which dissociates in water and yields hydrogen ions [H^+]. It is referred to as a proton donor.

Base – A base is a compound which dissociates in water and yields hydroxyl ions [OH^-]. Bases are proton acceptors.

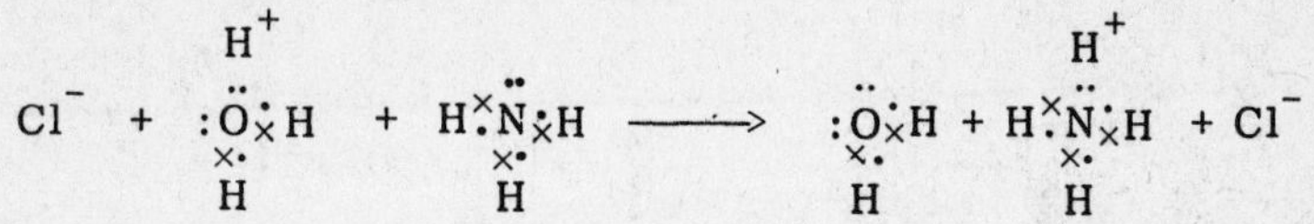

```
                H+                                              H+
          ..          ..                          ..     ..
Cl-  +  :O×H   +  H×N×H    ——>    :O×H + H×N×H  + Cl-
          ×.          ×.                          ×.     ×.
          H           H                           H      H
```

Reaction between hydrochloric acid (proton donor) and ammonia (proton acceptor).

pH – The degree of acidity or alkalinity is measured by pH.

$$pH = \frac{1}{\log[H^+]} = -\log[H^+]$$

pH = 7 → neutral

pH < 7 → acidic

pH > 7 → basic

Chemical Changes

Chemical Reaction – A chemical reaction is any process in which at least one bond is either broken or formed. The outcome of a chemical reaction is a rearrangement of atoms and bonding patterns.

Laws of Thermodynamics –

A) First Law of Thermodynamics – In any process, the sum of all energy changes must be zero.

B) Second Law of Thermodynamics – Any system tends toward a state of greater stability – stability meaning randomness, disorder, and probability.

C) Third Law of Thermodynamics – A perfect crystal which is a completely ordered system, at absolute zero (0^0 Kelvin) would have perfect order, and therefore its entropy would be zero.

Stability of chemical system depends on:

A) Enthalpy – Total energy content of a system.

B) Entropy – Energy distribution.

Exergonic Reaction – Exergonic reactions release free energy; all spontaneous reactions are exergonic.

Endergonic Reaction – Endergonic reactions require the addition of free energy from an external source.

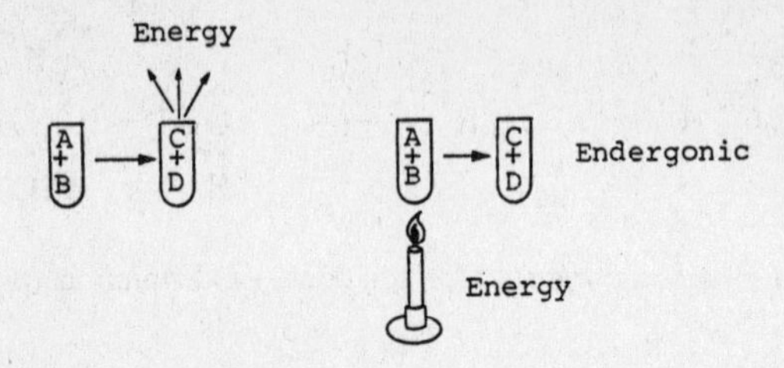

Organic Chemistry

Carbon is a unique element. It can bond with itself to form long chains with branches and rings. For this reason, it is the basis of all life on earth. Scientists have speculated whether life on other worlds could be based upon another element. The only other element that approaches carbon's ability to bond with itself is silicone. Silicone can form long chains with itself, but lacks the ability to form the useful branches that carbon does. Hence, it appears that life, as we know it, can only exist due to carbon. Because of its importance and unique bonding abilities, half of the study of chemistry is devoted to the study of carbon and its compounds. This branch of chemistry is known as Organic Chemistry.

Hydrocarbons – The simplest organic molecules are the hydrocarbons. These compounds are composed solely of carbon and hydrogen. They can exist as chains (e.g., butane) or rings (e.g., benzene).

```
    H   H   H   H
    |   |   |   |
H — C — C — C — C — H
    |   |   |   |
    H   H   H   H
```

butane

```
          H
          |
          C
       //   \
   H—C         C—H
     |         ||
   H—C         C—H
       \\   /
          C
          |
          H
```

benzene

A special branch of organic chemistry that deals with compounds and reactions important to living things (plant, animal, or microbial) is called Biochemistry. Today it is impossible to study biology without some knowledge of biochemistry.

Lipids – Lipids are organic compounds that dissolve poorly, if at all, in water (hydrophobic). All lipids (fats and oils) are composed of carbon, hydrogen, and oxygen where the ratio of hydrogen atoms to oxygen atoms is greater than 2:1. A lipid molecule is composed of 1 glycerol and 3 fatty acids.

Phospholipid – A phospholipid is a variety of a substituted lipid which contains a phosphate group.

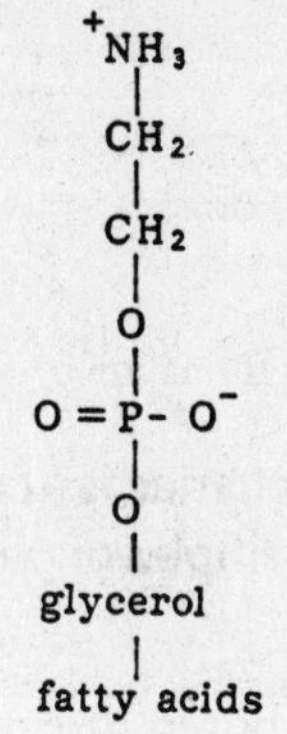

$^+NH_3$
|
CH_2
|
CH_2
|
O
|
$O = P - O^-$
|
O
|
glycerol
|
fatty acids

An example of the addition which makes a lipid a phospholipid.

Steroids – Steroids are complex molecules which contain carbon atoms arranged in four interlocking rings. Some steroids of biological importance are vitamin D, bile salts, and cholesterol.

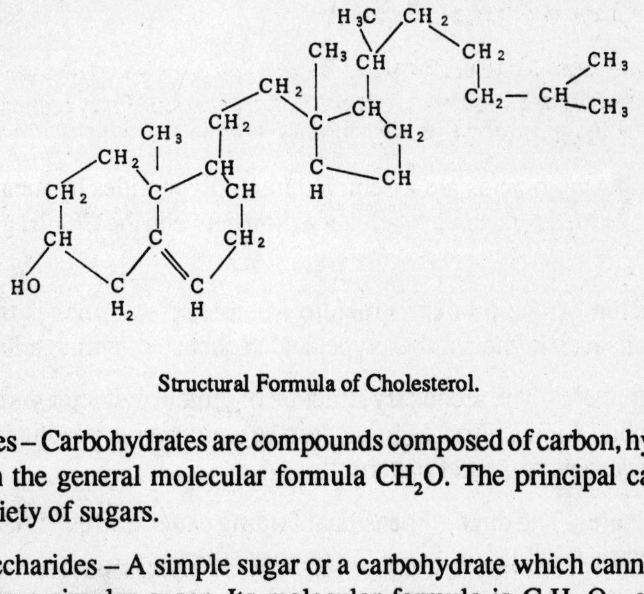

H_3C CH_2
CH_3 CH CH_2 CH_3
CH_2 CH_2 — CH CH_3
CH_3 CH_2 C CH CH_2
CH_2 CH C CH
CH_2 C CH H
CH C CH_2
HO C C
H_2 H

Structural Formula of Cholesterol.

Carbohydrates – Carbohydrates are compounds composed of carbon, hydrogen, and oxygen, with the general molecular formula CH_2O. The principal carbohydrates include a variety of sugars.

A) Monosaccharides – A simple sugar or a carbohydrate which cannot be broken down into a simpler sugar. Its molecular formula is $C_6H_{12}O_6$, and the most common is glucose.

OH — OH

Glucose.

B) Disaccharide – A double sugar or a combination of two simple sugar molecules. Sucrose is a familiar disaccharide as are maltose and lactose.

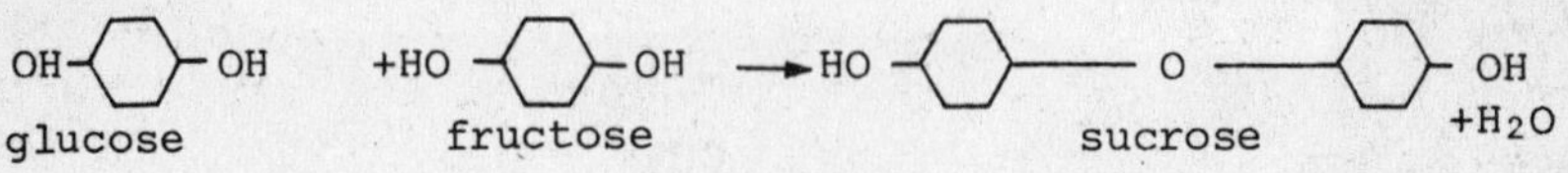

OH–⬡–OH (glucose) + HO–⬡–OH (fructose) → HO–⬡–O–⬡–OH (sucrose) + H_2O

Double sugar formation by dehydration synthesis.

C) Polysaccharide – A polysaccharide is a complex compound composed of a large number of glucose units. Examples of polysaccharides are starch, cellulose, and glycogen.

Proteins – All proteins are composed of carbon, hydrogen, oxygen, nitrogen, and sometimes phosphorus and sulfur. Approximately 50% of the dry weight of living matter is protein.

Amino Acids – The 20 amino acids are the building blocks of protein.

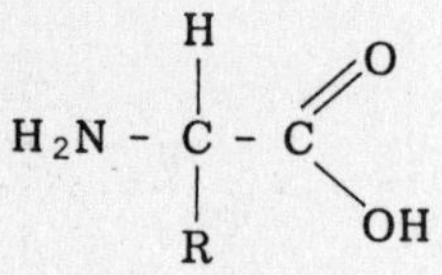

$$H_2N - \underset{\large R}{\overset{\large H}{C}} - C(=O)OH$$

An amino acid with R representing its distinctive side chain.

Polypeptides – Amino acids are assembled into polypeptides by means of peptide bonds. This is formed by a condensation reaction between the COOH groups and the NH_2 groups.

Primary Structure – The primary structure of protein molecules is the number of polypeptide chains and the number, type, and sequence of amino acids in each.

Secondary Structure – The secondary structure of protein molecules is characterized by the same bond angles repeated in successive amino acids which gives the linear molecule a recurrent structural pattern.

Tertiary Structure – The three-dimensional folding pattern, which is super-imposed on the secondary structure, is called the tertiary structure.

Quaternary Structure – The quaternary structure is the manner in which two or more independently folded subunits fit together.

Nucleic Acids – Nucleic acids are long polymers involved in heredity and in the manufacture of different kinds of proteins. The two most important nucleic acids are deoxyribonucleic acid (DNA) and ribonucleic acid (RNA).

Nucleotides – These are the building blocks of nucleic acids. Nucleotides are complex molecules composed of a nitrogenous base, a 5-carbon sugar and a phosphate group.

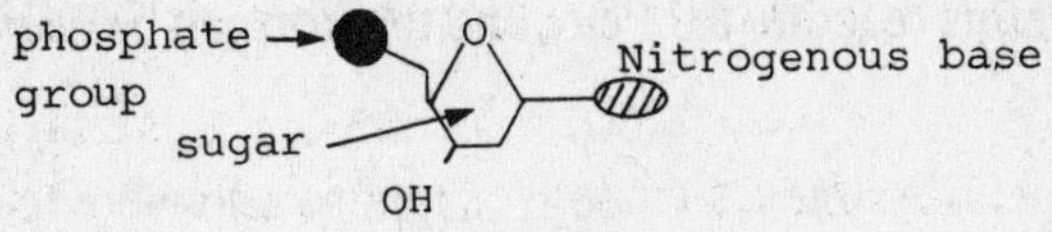

Structure of a nucleotide.

Deoxyribonucleic Acid (DNA) – Chromosomes and genes are composed mainly of DNA. It is composed of deoxyribose, nitrogenous bases, and phosphate groups.

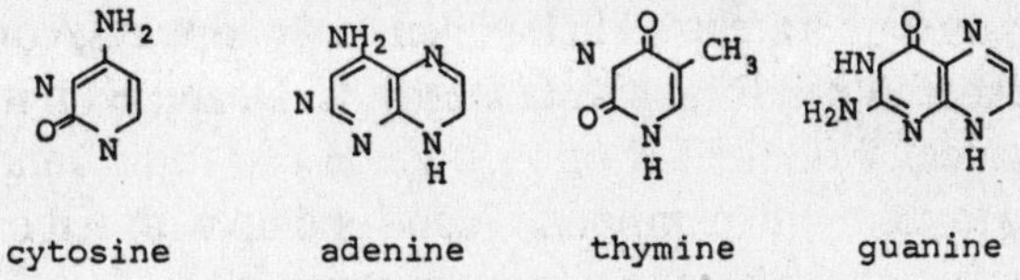

The four nitrogenous bases of DNA.

Ribonucleic Acid (RNA) – RNA is involved in protein synthesis. Unlike DNA, it is composed of the sugar ribose and the nitrogenous base uracil instead of thymine.

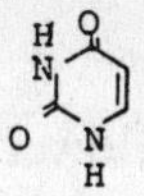

Uracil.

B. BIOCHEMICAL PATHWAYS

The above information is very necessary to understand metabolism. For example, how does the oxygen that an animal breathes affect its breakdown of glucose? Before we can begin to answer such questions, we must first address another class of biochemical compounds called enzymes. An enzyme is a special type of catalyst made of proteins. A catalyst is any substance that will lower the activation energy of a reaction so that it requires less energy to start and it will achieve equilibrium sooner. A catalyst will **not** cause a reaction to occur if it would not occur without the presence of the catalyst. During the reaction the catalyst is not used up, and hence, it can be used over and over again for the same reaction. Isaac Asimov makes the analogy of tying one's shoelaces. This can be done alone, but it is easier if one used a chair. In the process the chair is not used up, and another person can use the same chair to prop their foot up and tie their shoelace. In the same manner, hydrogen peroxide can break down to form water and oxygen gas ($2H_2O_2 \rightarrow 2H_2O + O_2\uparrow$), although this proceeds at a very low rate unassisted. However, if a few iron filings are dropped into the hydrogen peroxide, voluminous bubbles of oxygen will gush forth. The iron has acted as a catalyst, offering its sur– face as a place for this reaction to occur, and the iron can be used again and again for this reaction.

To continue with Asimov's analogy, suppose one were to make a chair tailor-made to aid shoelace tying. This would be able to accommodate even more people in the same amount of time. Thus it is with an enzyme. The enzyme called catalase is a protein that has a surface that is ideal for the decomposition of hydrogen peroxide, it works even better than iron.

To see how enzymes are used in biochemical pathways, consider the following situation: substance A can form substances B, C, D, and E on its own. Substance B can form substances F, G, and H on its own. H can form I and J unassisted. If an organism needs to start with compound A and end up with J, it can do so through the following steps using enzymes to direct the synthesis:

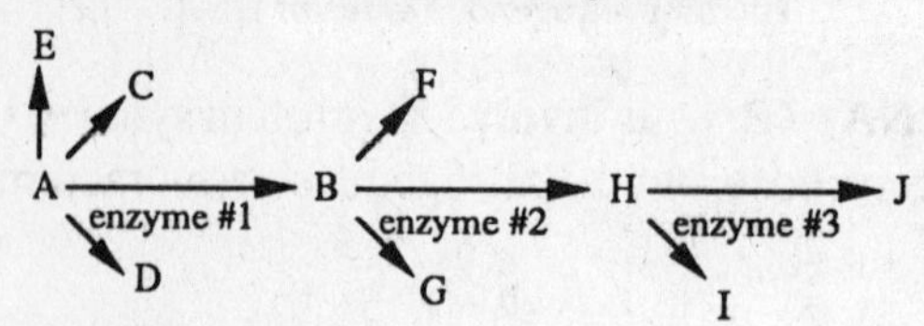

Enzymes are protein catalysts that lower the amount of activation energy needed for a reaction, allowing it to occur more rapidly. The enzyme binds with the substrate but resumes its original conformation after forming the enzyme-substrate complex.

Substrate – A substrate is the molecule upon which an enzyme acts. The atoms of the substrate are arranged so as to fit into the active site of the enzyme that acts upon it. An enzyme usually affects only one substrate.

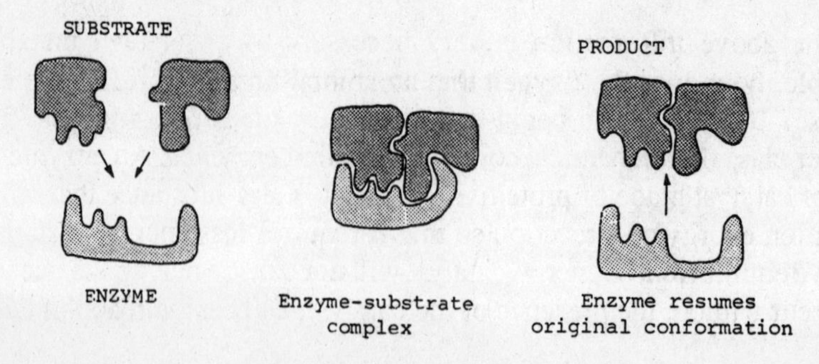

The action of an enzyme.

Allosteric Enzyme – An allosteric enzyme is one that can exist in two distinct conformations. Usually, the enzyme is active in one conformation and inactive in the other.

Allosteric Inhibition – During the process of allosteric inhibition, an inhibitory molecule called a negative modulator binds to the enzyme and stabilizes it in its inactive conformation.

Coenzymes – These are metal ions or non-proteinaceous organic molecules that bind briefly and loosely to some enzymes. The coenzyme is necessary for the catalytic reaction of such enzymes.

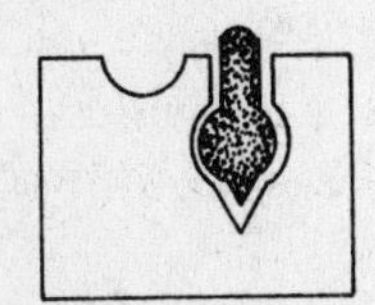

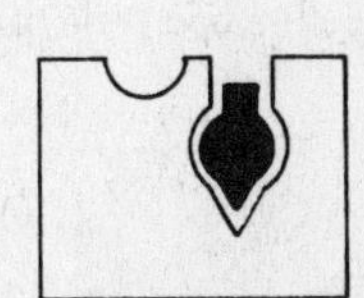

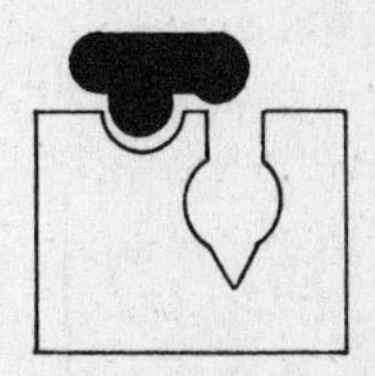

A: enzyme-substrate complex; B: competitive inhibition; C: non-competitive inhibition.

Factors influencing the rate of enzyme action:

1. pH

2. temperature

3. concentration of enzyme and substrate

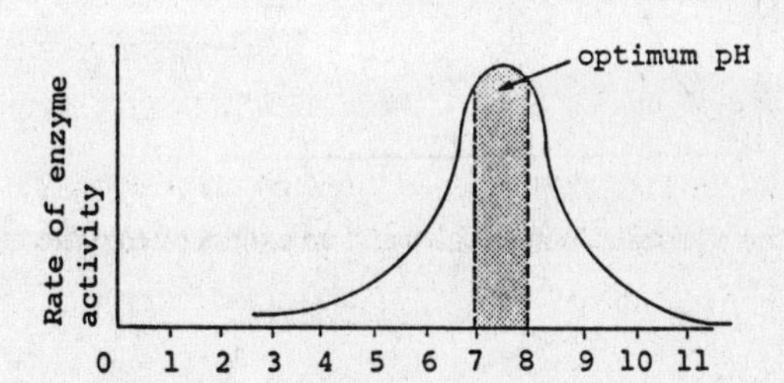

Effect of pH on rate of enzyme action.

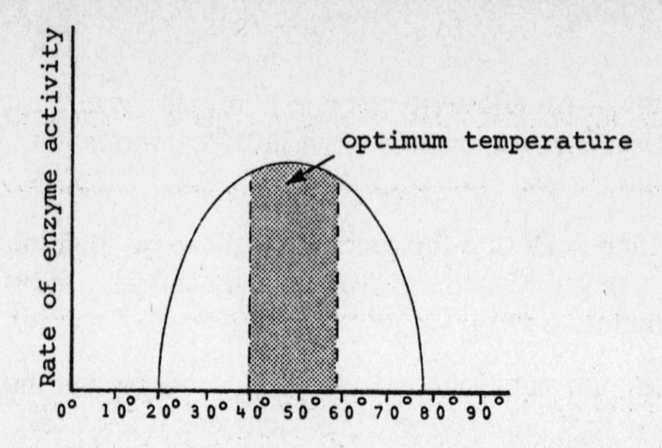

Effect of temperature upon rate of enzyme activity.

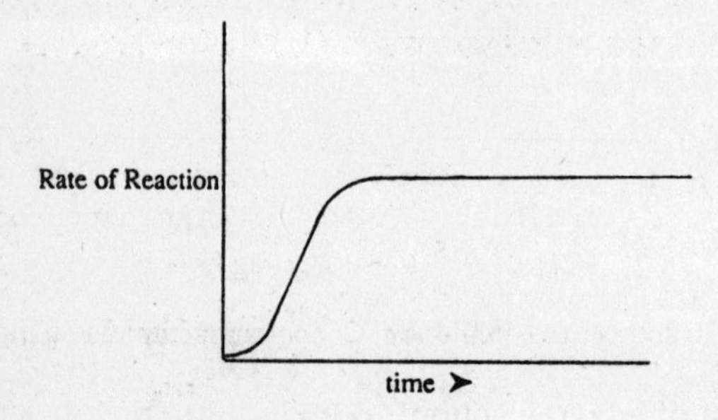

Fixed amount of enzyme and an excess of substrate molecules.

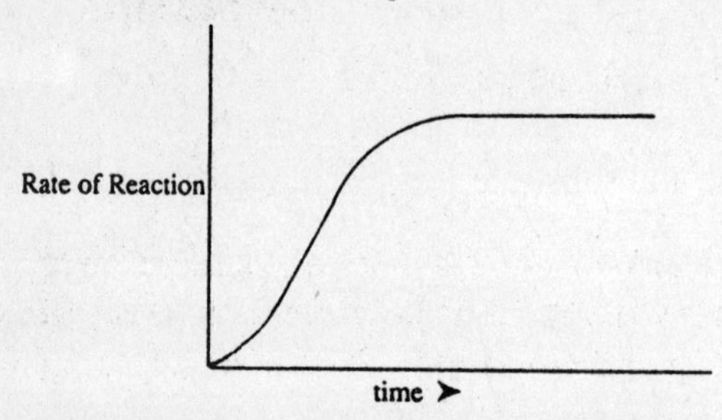

Fixed number of substrate molecules and an excess of enzyme molecules.

Photosynthesis

Photosynthesis is the basic food-making process through which inorganic CO_2 and H_2O are transformed to organic compounds, specifically carbohydrates.

Chloroplasts absorb light energy and use CO_2 and H_2O to synthesize carbohydrates. Oxygen, which is formed as a by-product, is either eliminated into the air through the stomates, stored temporarily in the air spaces, or used in cellular respiration.

An overall chemical description of photosynthesis is the equation

$$6\ CO_2 + 6\ H_2O \xrightarrow[\text{chlorophyll}]{\text{light}} C_6H_{12}O_6 + 6\ O_2$$

Light Reaction (Photolysis) – A first step in photosynthesis is the decomposition of water molecules to separate hydrogen and oxygen components. This decomposition is associated with processes involving chlorophyll and light and is thus known as the light reaction.

Dark Reaction (CO_2 Fixation) – In this second phase, the hydrogen that results from photolysis reacts with CO_2 and carbohydrate forms. CO_2 fixation does not require light.

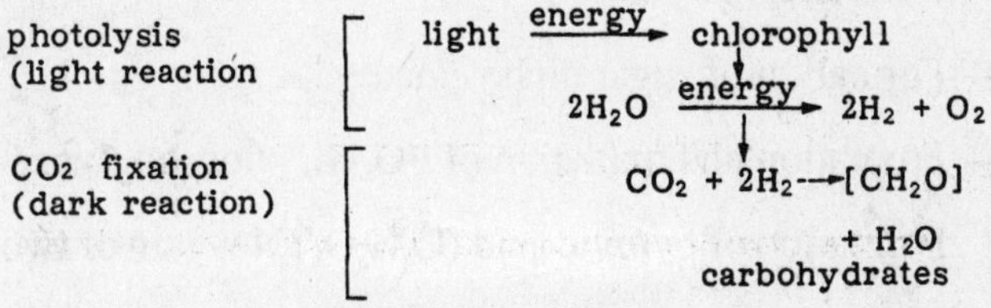

photolysis (light reaction): light $\xrightarrow{\text{energy}}$ chlorophyll; $2H_2O \xrightarrow{\text{energy}} 2H_2 + O_2$

CO_2 fixation (dark reaction): $CO_2 + 2H_2 \rightarrow [CH_2O] + H_2O$ carbohydrates

Photolysis and CO_2 fixation.

Cellular Respiration

During cellular respiration, glucose must first be activated before it can break down and release energy. After activation, glucose enters into numerous reactions occurring in two stages. One stage is the anaerobic phase of cellular respiration and the other is the aerobic phase.

Glycolysis – Glycolysis refers to the breakdown of glucose which marks the start of the anaerobic reactions of cellular respiration. ATP is the energy source which activates glucose and initiates the process of glycolysis.

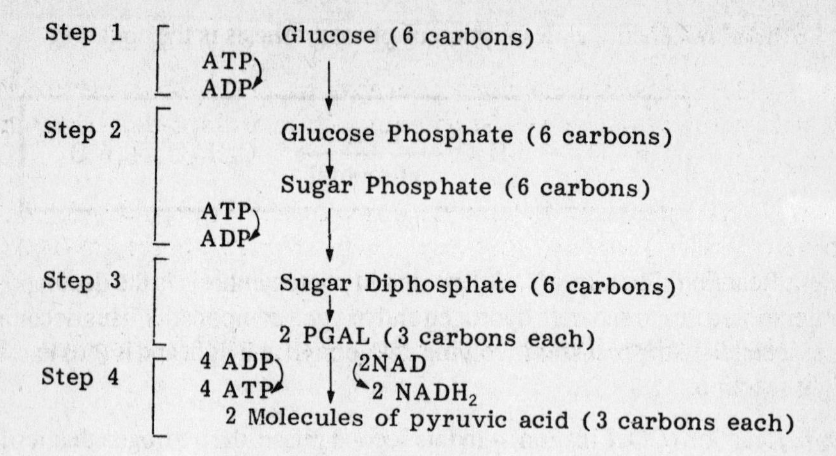

The major steps in glycolysis.

The steps in the figure above are summarized as follows:

Step 1 – Activation of glucose

Step 2 – Formation of sugar diphosphate

Step 3 – Formation and oxidation of PGAL, phosphoglyceraldehyde

Step 4 – Formation of pyruvic acid ($C_3H_4O_3$) Net gain of two ATP molecules

Krebs Cycle (Citric Acid Cycle) – The Krebs Cycle is the final common pathway by which the carbon chains of amino acids, fatty acids, and carbohydrates are metabolized to yield CO_2. Pyruvic acid is converted to acetyl coenzyme A and, through a series of reactions, citric acid is formed.

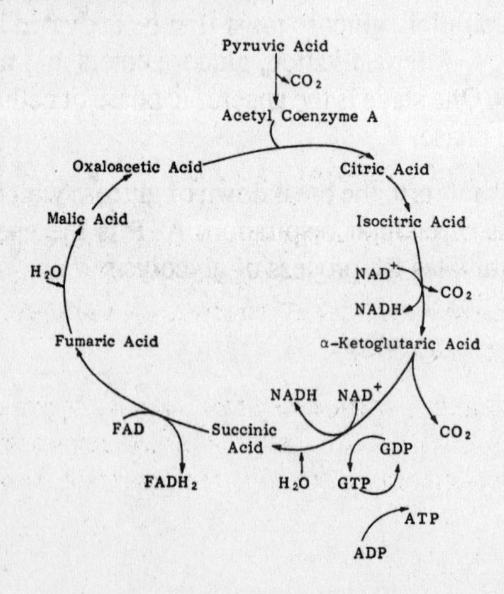

Summary of the Krebs Cycle.

ATP and NAD

ATP – ATP stands for adenosine triphosphate and is a coenzyme essential for the breakdown of glucose. When the bonds in ATP are hydrolyzed, a large amount of energy is released.

NAD – NAD stands for nicotinamide adenine dinucleotide. Like ATP, it is also a coenzyme. NAD participates in a large number of oxidation-reduction reactions in cells, including those in cellular respiration.

The Respiratory Chain (Hydrogen Transport System)

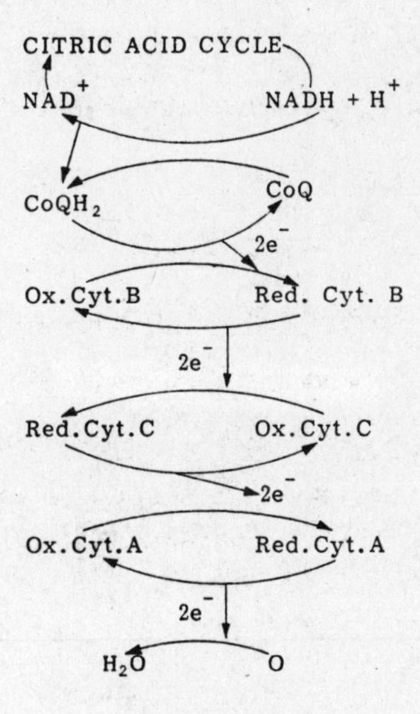

The Respiratory Chain.

This system consists of a series of enzymes and coenzymes which pick-up, hold, and then transfer hydrogen atoms among themselves until the hydrogen reaches its final acceptor, which is oxygen. Cytochromes are the enzymes and coenzymes involved in transferring hydrogen.

The cytochromes, together with other enzymes, split hydrogen atoms attached to compounds such as $NADH_2$, into hydrogen ions and electrons. Each cytochrome then passes the hydrogen ions and electrons to another cytochrome in the series.

C. MOLECULAR GENETICS

Deoxyribonucleic acid, or DNA, is referred to as "The Blueprints of Life." By this statement, it is meant that DNA is what determines what a developing organism will be, a horse, a radish, or a mushroom, for example. DNA also helps to determine minor differences among the same species, such as hair color or leaf size.

In order to do the above, DNA must first be able to replicate itself faithfully, to be able to pass this information on to future generations. Second, there must be a faithful method to read these "blueprints" and express them.

First we shall start with the structure of DNA:

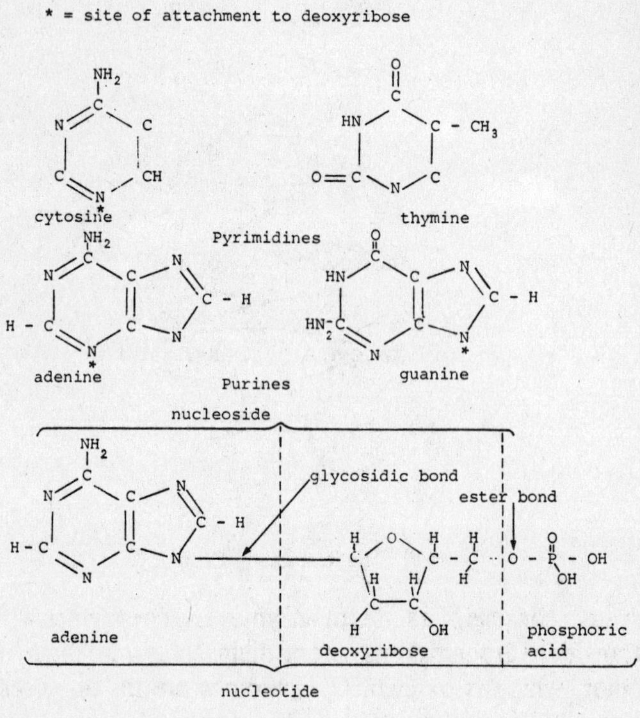

* = site of attachment to deoxyribose

cytosine

thymine

Pyrimidines

adenine

guanine

Purines

nucleoside

glycosidic bond

ester bond

adenine

deoxyribose

phosphoric acid

nucleotide

Structural formulas of purines (adenine and guanine), pyrimidines (thymine and cytosine), and a nucleotide.

Deoxyribonucleic acid is made up of a nitrogenous base, a five carbon sugar (deoxyribose) and phosphate groups. DNA may contain one of four nitrogenous bases which are the purines, adenine and guanine, and the pyrimidines, cytosine and thymine. Each nitrogenous base is attached to deoxyribose via a glycosidic linkage, and deoxyribose is attached to the phosphate group by an ester bond. This base sugar-phosphate combination is a nucleotide. There are four kinds of nucleotides in DNA, each containing one of the four nitrogenous bases. The nucleotides are joined by phosphate ester bonds into a chain. DNA is made up of two complementary chains of nucleotides.

The structure of DNA:

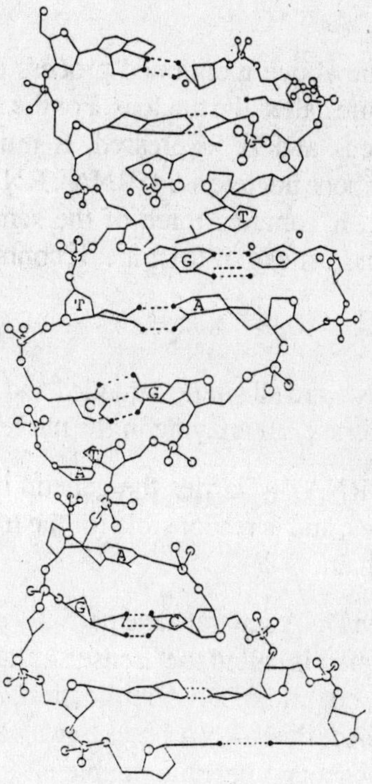

The DNA double helix. The purine (G=Guanine, A=Adenine) and pyrimidine (T=Thymine, C=Cytosine) base pairs are connected by hydrogen bonds. When these bonds are broken, the DNA can unwind and replicate (as in mitosis) or act as a template for mRNA synthesis. The ribose sugar and phosphate groups act as a backbone, helping to maintain the sequential order of the nitrogenous bases. There are ten base pairs in one turn.

In 1953, Watson and Crick proposed a model of the DNA molecule which consisted of a double helix of nucleotides in which the two nucleotide helices were wound around each other. Each full turn of the helix measure 34 Å, and contains 10 nucleotides equally spaced from each other. The radius of the helix is 10 Å. These measurements were in agreement with those obtained from x-ray diffraction patterns of DNA.

Remember that when replicating, adenine (A) always pairs with thymine (T), and cytosine (C) always pairs with guanine (G). In short, if a strand of DNA were to be uncoiled and one side were found to read: A C T T A G, it would be easy
| | | | | |
enough to determine the opposite side, and hence the complete strand as:

```
A C T T A G
| | | | | |
T G A A T C
```

Gene Expression: The above mentioned process of DNA replication usually occurs in a cellular structure called the nucleus. For the information contained in the DNA to leave the nucleus and be expressed, it must take on the form of an intermediate substance, ribose nucleic acid (RNA). RNA is much like DNA, except that it is single-stranded. It contains three of the same bases as DNA (adenine, cytosine, and guanine) and instead of thymine, it contains uracil (U).

Types of RNA

The three types of RNA are all single-stranded and are transcribed from a DNA template by RNA polymerase, an enzyme in the nucleus.

A) Messenger RNA (mRNA) – Carries the genetic information coded for in the DNA to the ribosomes, and is responsible for the translation of that information into a polypeptide chain.

The two strands of the DNA separate and partially uncoil. The mRNA is copied. The strand that is copied is called the "sense strand." As in DNA replication, C always pairs with G, but unlike DNA replication, A pairs with U, since there is no T. After it is formed, the mRNA breaks away from the DNA template, and the DNA strands once again become double stranded and coil up. The newly formed RNA strand leaves the nucleus, and travels into the cytoplasm (the area in the cell outside of the nucleus). The entire process just mentioned of converting the genetic message from the DNA into RNA is called transcription.

Expressing the genetic message contained in the RNA, i.e. building the protein it specifies is called translation. Once in the cytoplasm, the mRNA strand connects to a structure called a ribosome. Here the message contained in the mRNA is translated into protein. The order of the bases in the mRNA determines the order that the ribosome will connect amino acids. (A protein is built of amino acids). Every three RNA bases specify a particular amino acid.

For example, using the chart below, one can see the UUC codes for Phenylalanine (Phe), UAC tyrosine (Tyr), and CUC leucine (Leu). Note that some amino acids are coded for by more than one group of codons. Also note that some combinations do not code for any amino acids, hence they terminate the production of the protein. Those marked with an NH_2 initiate protein synthesis. The protein thus produced may be either a fibrous, or structural type (e.g., hair) or a globular, or enzyme type (e.g., hemoglobin).

The Genetic Code

First Position (5′ end)	Second position	Third position (3′ end) U	C	A	G
U	U	Phe	Phe	Leu	Leu
	C	Ser	Ser	Ser	Ser
	A	Tyr	Tyr	Terminator	Terminator
	G	Cys	Cys	Terminator	Trp
C	U	Leu	Leu	Leu	Leu
	C	Pro	Pro	Pro	Pro
	A	His	His	Glu NH_2	Glu NH_2
	G	Arg	Arg	Arg	Arg
A	U	Ileu	Ileu	Ileu	Met
	C	Thr	Thr	Thr	Thr
	A	Asp NH_2	Asp NH_2	Lys	Lys
	G	Ser	Ser	Arg	Arg
G	U	Val	Val	Val	Val
	C	Ala	Ala	Ala	Ala
	A	Asp	Asp	Asp	Asp
	G	Gly	Gly	Gly	Gly

B) Transfer RNA (tRNA) – Moves the amino acids into the proper positions that the mRNA calls for.

This is done by the tRNA for a specific amino acid having the anti codon for that acid. For example, looking at the table above, one can see that valine can be coded for by the codon GUG; therefore, the anti codon will be CAC (A to U and C to G). A tRNA that carries the CAC anti codon would then also be attached to valine. Below is an illustration of the configuration of a tRNA for alanine. It should be kept in mind that this is a two-dimensional representation of a three-dimensional structure.

C) Ribosomal RNA (rRNA) – Ribosomal RNA is an important part of the ribosome where the proteins are made.

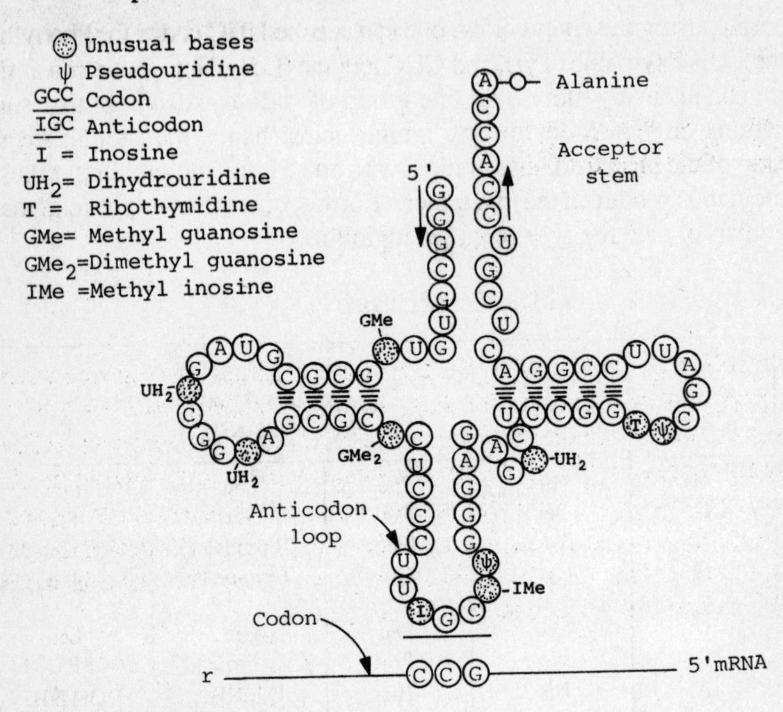

The complete nucleotide sequence of alanine tRNA showing the unusual bases and codon/anticodon position. Structure shown is two-dimensional.

CHAPTER 2

THE CELL

The basic unit of life is the cell. Cells are basically small bags that contain the biologically important molecules of life (e.g., nucleic acids, enzymes, etc.). These bags are dynamic, in that smaller molecules are constantly reentering and exiting them at a highly regulated rate, and more molecules are constantly being created inside of these bags, again, under constant regulation. Most important of all, these bags that we call cells can divide and form new cells. All living cells come from pre-existing cells. There are two main divisions of the types of cells: **Prokaryotic**, a more primitive type, and **Eukaryotic**, a more advanced type.

A. PROKARYOTIC CELLS

Prokaryotic cells are quite small and simple. It is thought that they were the first type of cell to appear on the Earth. Prokaryotic cells are usually quite small in size, usually only a couple of micrometers in diameter. No structures can be seen inside prokaryotic cells using a light microscope. The liquid contents of any cell (whether prokaryotic or eukaryotic) is referred to by the generic term cytoplasm. Prokaryotes have such small cellular volume that their cytoplasm has no where to go, and hence does not move around within the cell. There is a cellular membrane that marks the barrier between the inside of the cell and its external environment. Most prokaryotic cells also have a cell wall surrounding the outside of their cellular membrane. Also present, but invisible to the light microscope, are ribosomes, and a single circular chromosome made of DNA without any associated protein. It is this single molecule that regulates the cell's activities. Today, the only living prokaryotic cells are the bacteria and blue-green algae.

B. EUKARYOTIC CELLS

Eukaryotic cells are much larger and more complex than prokaryotic cells. The DNA is associated with specific proteins called "histones." Together, these make linear chromosomes that can be seen easily with a light microscope during certain times in the cell's cycle. The number of chromosomes present depends upon the type of organism the cell is or comes from. In eukaryotic cells the chromosomes are found in a special structure called the nucleus. Eukaryotic cells are so large that they usually contain many small structures, or organelles, as they are called. The cells are so large that the cytoplasm can be seen flowing from one location within the cell to another. This phenomenon is termed "cytoplasmic streaming." The cytoplasm when found in the nucleus is termed "nucleoplasm." A more general term that does not reveal location is "protoplasm." Below is a list of organelles that might (but not necessarily must) be found in eukaryotic cells.

Cellular Organelles

Mitochondria – Mitochondria are organelles with double-membranes that are the site of chemical reactions that extract energy from foodstuffs and make it available to the cell for all of its energy demanding activities.

Chloroplasts – These are found only in the cells of plants and certain algae. Photosynthesis occurs in the chloroplasts.

Plastids – These structures are present only in the cytoplasm of plant cells. The most important plastid, chloroplast, contains chlorophyll, a green pigment.

Lysosomes – Lysosomes are membrane-enclosed bodies that function as storage vesicles for many digestive enzymes.

Endoplasmic Reticulum – The endoplasmic reticulum (ER) transports substances within the cell.

Ribosomes – These organelles are small particles composed chiefly of ribosomal-RNA and are the sites of protein synthesis.

Golgi Apparatus – The functions of the Golgi apparatus include storage, modification, and packaging of secretory products.

Peroxisomes – Peroxisomes are membrane-bounded organelles which contain powerful oxidative enzymes.

Vacuoles – Vacuoles are membrane-enclosed, fluid-filled spaces. They have their greatest development in plant cells where they store materials such as soluble organic nitrogen compounds, sugars, various organic acids, some proteins, and several pigments.

Cell Wall – This is only present in plant cells and is used for protection and support.

Centriole and Centrosome – These function in cell division. They are present only in animal cells.

Cilia and Flagella – These are hairlike extensions from the cytoplasm of a cell. They both show coordinated beating movements, which are the major means of locomotion and ingestion in unicellular organisms.

Nucleolus – The nucleolus is a generally oval body composed of protein and RNA. Nucleoli are produced by chromosomes and participate in the process of protein synthesis.

Below is a diagram of a "typical animal cell" and a "typical plant cell." These diagrams are not typical in that a real cross-section of a cell will never show all of these organelles at the same time.

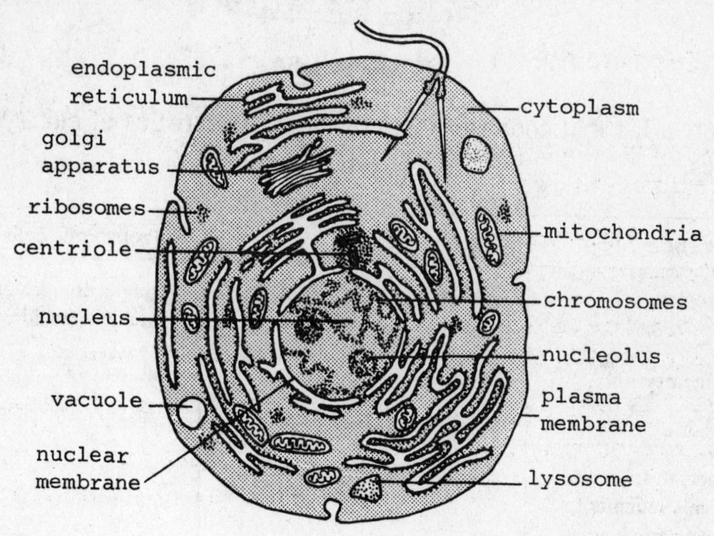

Typical animal cell.

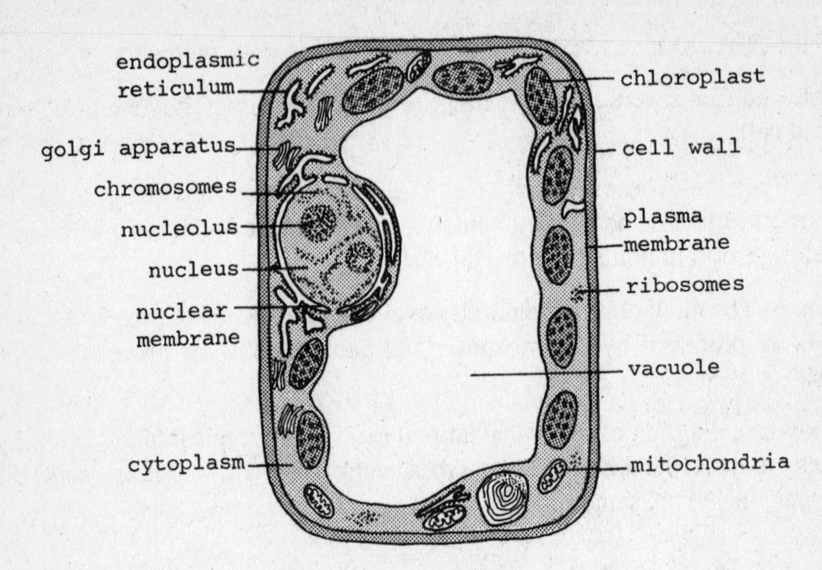

Typical plant cell.

Below is a list that contrasts some of the characteristics of eukaryotics and prokaryotics.

Characteristic	Eukaryotic cells	Prokaryotic cells
Chromosomes	multiple, composed of nucleic acids and protein	single, composed only of nucleic acid
Nuclear membrane	present	absent
Mitochondria	present	absent
Golgi apparatus, endoplasmic reticulum, lysosomes, peroxisomes	present	absent
Photosynthetic apparatus	chlorophyll, when present, is contained (in chloroplasts)	may contain chlorophyll
Microtubules	present	rarely present
Ribosomes	large	small
Flagella	have 9-2 tubular structure	lack 9-2 tubular structure
Cell wall	when present, does not contain muramic acid	contains muramic acid

C. CELLULAR DIVISION

As was mentioned earlier, all cells come from pre-existing cells. Prokaryotic cells do this by a simple method called "binary fission," where a cell experiences a simple invagination of the cell membrane, followed by the cell wall, resulting eventually in a membrane with a new cell wall forming, dividing the cell into two (usually equal) halves as can be seen in the diagram below:

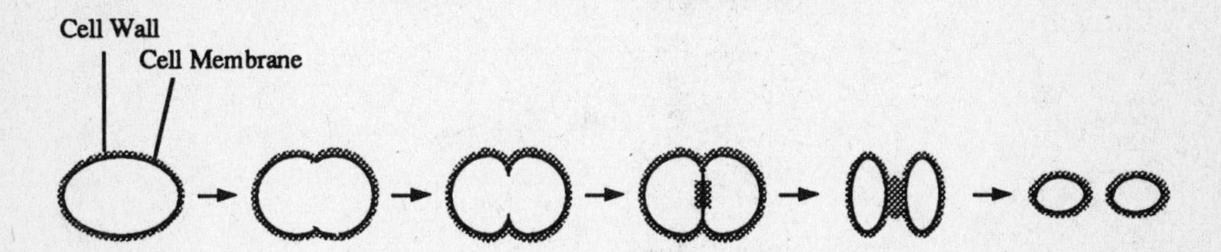

This process can take from as little as 20 minutes to a few hours. Eukaryotes, being more complex, divide by going through a very complex process consisting of several definite stages, known as "mitosis." This process can take place within a few hours to many years. Below is a list of the stages found in mitosis:

Mitosis – Mitosis is a form of cell division whereby each of two daughter nuclei receives the same chromosome complement as the parent nucleus. All kinds of asexual reproduction are carried out by mitosis; it is also responsible for growth, regeneration, and cell replacement in multicellular organisms.

A) Interphase – Interphase is no longer called the resting phase because a great deal of activity occurs during this phase. In the cytoplasm, oxidation and synthesis reactions take place. In the nucleus, DNA replicates itself and forms messenger RNA, transfer RNA, and ribosomal RNA.

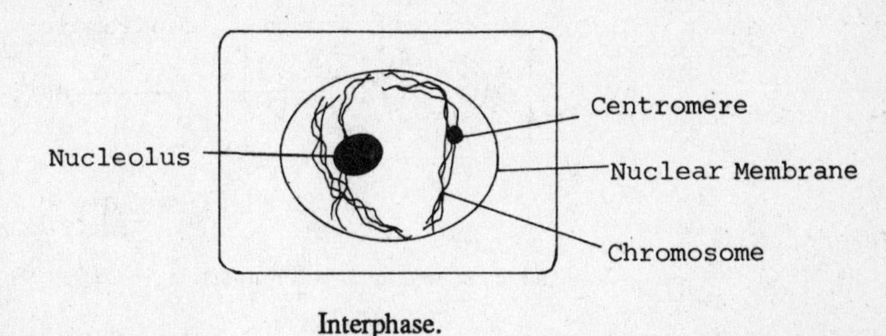

Interphase.

B) Prophase – Chromatids shorten and thicken during this stage of mitosis. The nucleoli disappear and the nuclear membrane breaks down and disappears as well. Spindle fibers begin to form. In an animal cell, there is also division of the centrosome and centrioles.

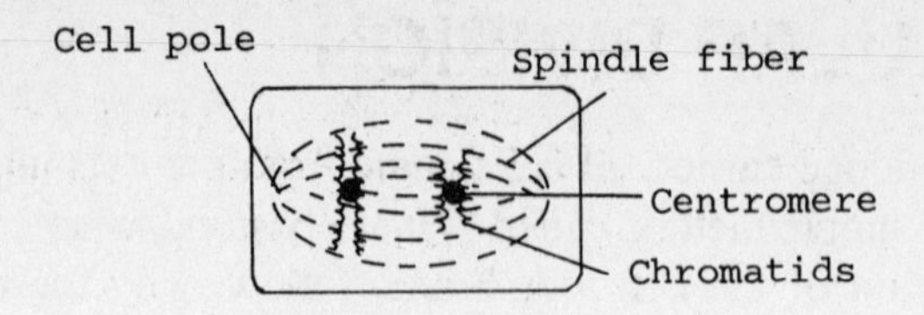

Late prophase in plant cell mitosis.

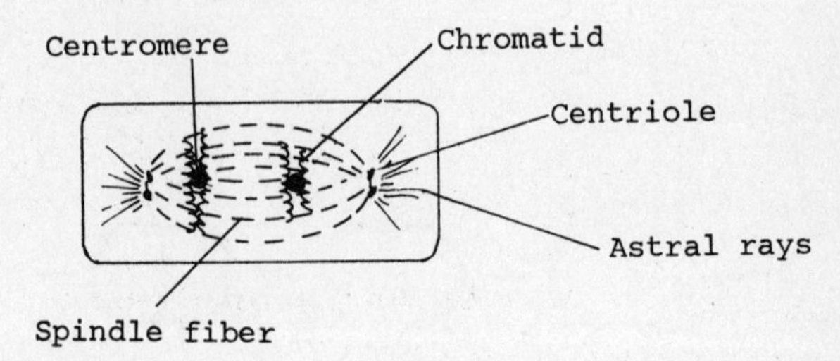

Prophase in animal cell mitosis.

C) Metaphase – During this phase, each chromosome moves to the equator, or middle of the spindle. The paired chromosomes attach to the spindle at the centromere.

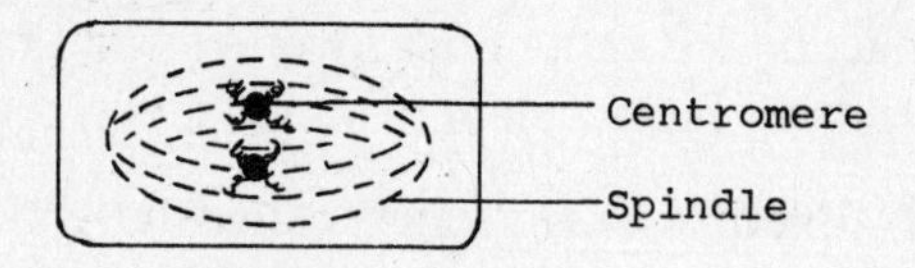

Metaphase in plant cell mitosis.

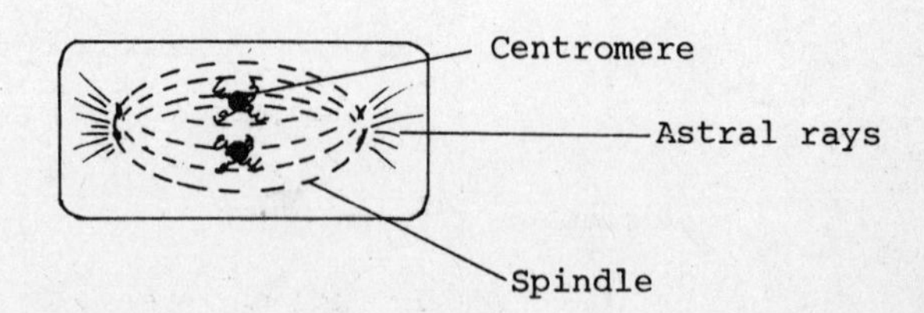

Metaphase in animal cell mitosis.

D) Anaphase – Anaphase is characterized by the separation of sister chromatids into a single-stranded chromosome. The chromosomes migrate to opposite poles of the cell.

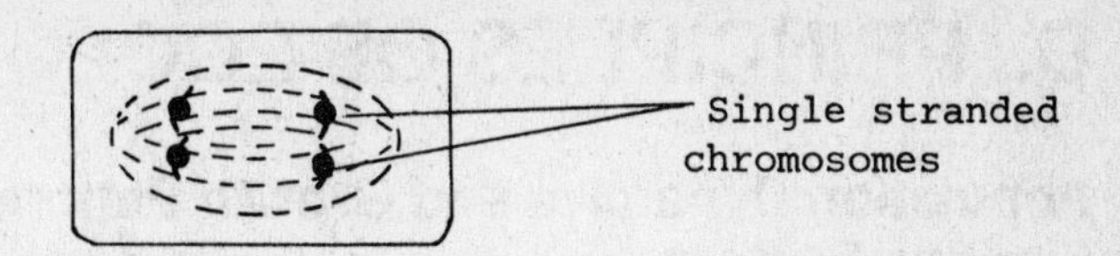

Anaphase in plant cell mitosis.

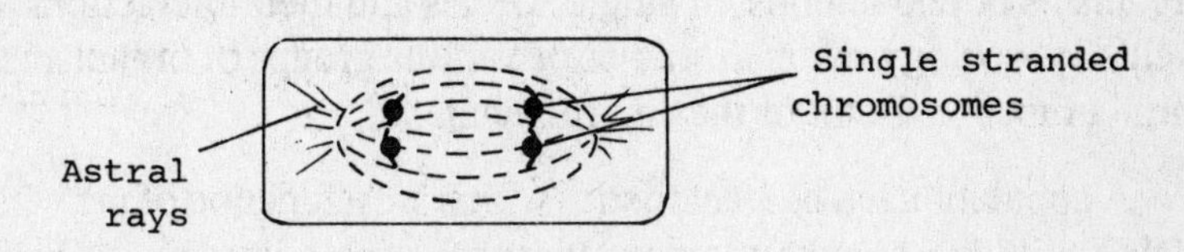

Anaphase in animal cell mitosis.

E) Telophase – During telophase, the chromosomes begin to uncoil and the nucleoli as well as the nuclear membrane reappear. In plant cells, a cell plate appears at the equator which divides the parent cell into two daughter cells. In animal cells, an invagination of the plasma membrane divides the parent cell.

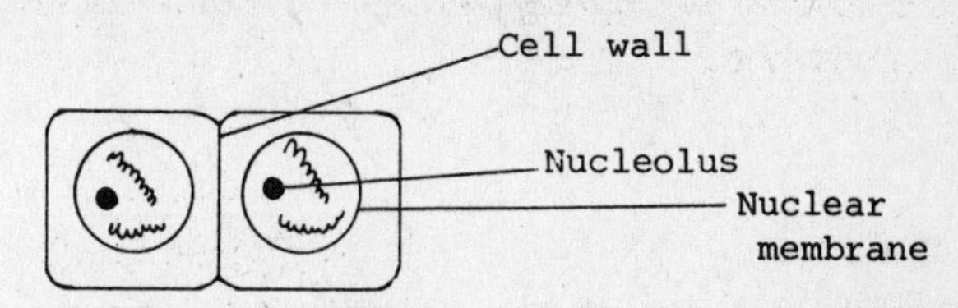

Late telophase in animal cell.

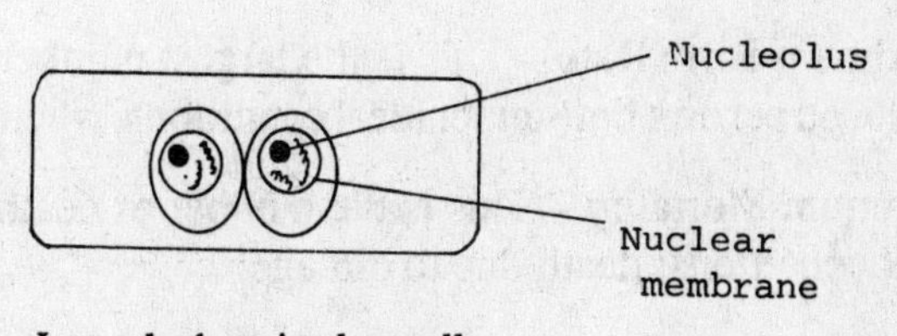

Late telophase in plant cell.

D. PRINCIPLES OF ECOLOGY

Population Dynamics and Growth Patterns, Biotic Potential, Limiting Factors

Ecology can be defined as the study of the interactions between groups of organisms and their environment. The term autecology refers to studies of individual organisms or populations, of single species and their interactions with the environment. Synecology refers to studies of various groups of organisms that associate to form a functional unit of the environment.

A population has characteristics which are a function of the whole group and not of the individual members; among these are population density, birth rate, death rate, age distribution, and biotic potential.

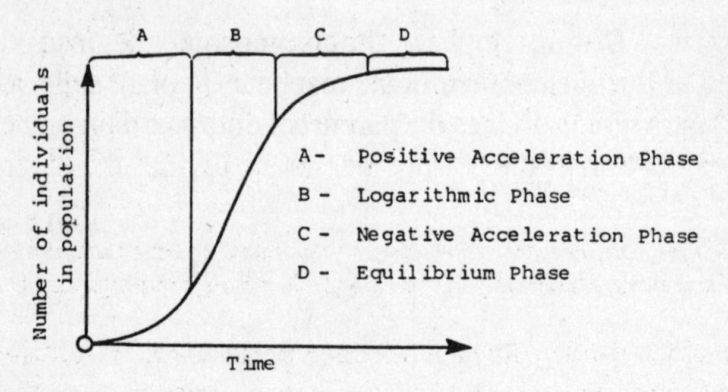

A typical S-shaped growth curve of a population in which the total number of individuals is plotted against time.

A) Population Density – The number or mass of individuals per area or volume of habitable space.

B) Maximum Birth Rate – This is the largest number of organisms that could be produced per unit time under ideal conditions, when there are no limiting factors.

C) Minimum Mortality – This is the number of deaths which would occur under ideal conditions; death due to old age.

D) Biotic Potential (reproductive potential) – The ability of a population to increase in numbers when the age ratio is stable and all environmental conditions are optimal.

Ecosystems and Communities: Energetics and Energy Flow, Productivity, Species Interactions, Succession

The energy cycle starts with sunlight being utilized by green plants on earth. The kinetic energy of sunlight is transformed into potential energy stored in chemical bonds in green plants. The potential energy is released in cell respiration and is used in various ways.

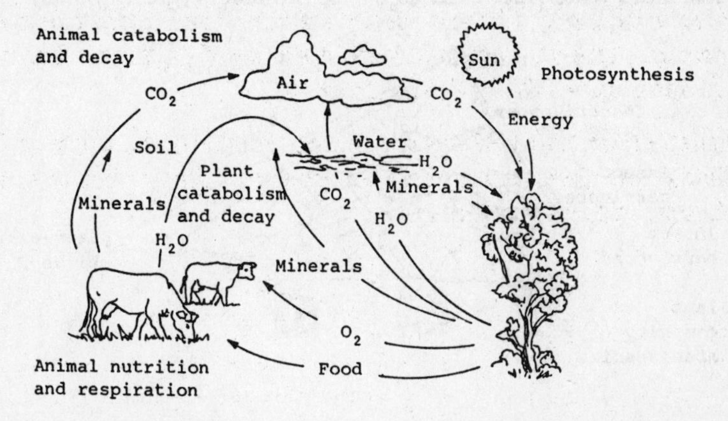

The energy cycle. This diagram shows the relationships between plants and animals and the nonliving materials of the earth. The energy of the sunlight is the only thing that is not returned to its source.

Some of the food synthesized by green plants are broken down by the plants for energy, releasing carbon dioxide and water. Bacteria and fungi break down the bodies of dead plants, using the liberated energy for their own metabolism. Carbon dioxide and water are then released and recycled.

Species interact in the following ways:

A) Contest competition is the active physical confrontation between two organisms which allows one to win the resources.

B) Scramble competition is the exploitation of a common vital resource by both species.

C) Mutualism is a type of relationship where both species benefit from one another. An example is nitrogen-fixing bacteria that live in nodules in the roots of legumes.

D) Commensalism is a relationship between two species in which one specie benefits while the other receives neither benefit nor harm; for example, epiphytes grow on the branches of forest trees.

E) Parasitism is a relationship where the host organism is harmed. Parasites can be classified as external or internal.

F) Succession is a fairly orderly process of changes of communities in a region. It involves replacement of the dominant species within a given area by other species.

Every food web begins with the autotrophic organisms (mainly green plants) being eaten by a consumer. The food web ends with decomposers, the organisms of decay, which are bacteria and fungi that degrade complex organic materials into simple substances which are reusable by the producers (green plants).

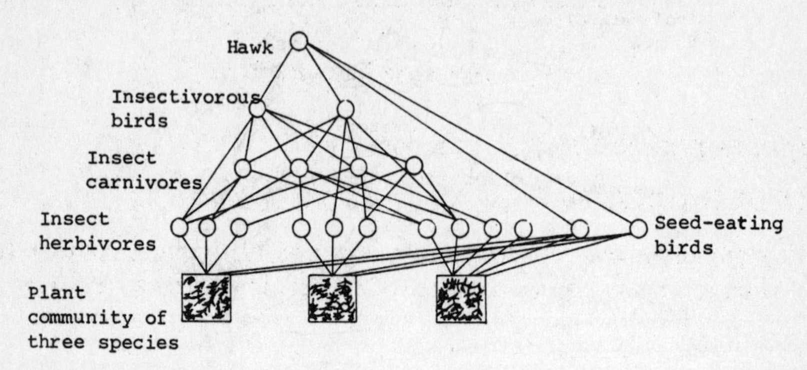

Hypothetical food web. It is assumed that there are 300 species of plants, 10 species of insect herbivores, two bird herbivores, two bird insectivores, and one hawk. In a real community, there would not only be more species at each trophic level, but also many animals that feed at more than one level, or that change levels as they grow older. Some general conclusions emerge from even an oversimplified model like this, however. There is an initial diversity introduced by the number of plants. This diversity is multiplied at the plant-eating level. At each subsequent level the diversity is reduced as the food chains converge.

Herbivores consume green plants, and may be acted upon directly by the decomposers or fed upon by secondary consumers, the carnivores. The successive levels in the food webs of a community are referred to as trophic levels.

A pyramid with the producers at the base, and the primary consumers at the apex, can show how energy is being supplied by the producers. It can also show a decrease of energy from the base of the apex accompanied by a decrease in numbers of organisms.

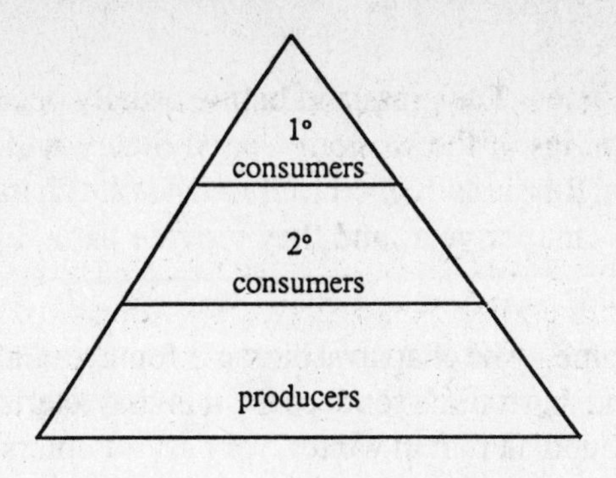

Pyramid of energy and numbers.

Biosphere and Biomes

TYPES OF HABITATS

There are four major habitats: marine, estuarine, fresh water, and terrestrial. A biome is a large community characterized by the kinds of plants and animals present.

Types of Biomes:

A) The Tundra Biome – A tundra is a band of treeless, wet, arctic grassland stretching between the Arctic Ocean and polar ice caps and the forests to the south. The main characteristics of the tundra are low temperatures and a short growing season.

B) The Forest Biomes

1) The northern coniferous forest stretches across North America and Eurasia just south of the tundra. The forest is characterized by spruce, fir, and pine trees and by such animals as the wolf, the lynx, and the snowshoe hare.

2) The moist coniferous forest biome stretches along the west coast of North America from Alaska south to central California. It is characterized by great humidity, high temperatures, high rainfall, and small seasonal ranges.

3) The temperature deciduous forest biome was found originally in eastern North America, Europe, parts of Japan and Australia, and the southern part of South America. It is characterized by moderate temperatures with distinct summers and winters and abundant, evenly distributed rainfall. Most of this forest region has now been replaced by cultivated fields and cities.

4) The tropical rain forests stretch around low lying areas near the equator. Dense vegetation, annual rainfall of 200 cm. or more, and a tremendous variety of animals characterize this area.

C) The Grassland Biome – The grassland biome usually occupies the interiors of continents, the prairies of the western United States, and those of Argentina, Australia, southern Russia and Siberia. Grasslands are characterized by rainfalls of about 25 to 75 cm. per year, and they provide natural pastures for grazing animals.

D) The Chaparral Biome – The chaparral biome is found in California, Mexico, the Mediterranean, and Australia's south coast. It is characterized by mild temperatures, relatively abundant rain in winter, very dry summers and trees with hard, thick evergreen leaves.

E) The Desert Biome – The desert is characterized by rainfall of less than 25 cm. per year, and sparse vegetation that consists of greasewood, sagebrush, and cactus. Such animals as the kangaroo rat and the pocket mouse are able to live there.

F) The Marine Biome – Although the saltiness of the open ocean is relatively uniform, the concentration of phosphates, nitrates, and other nutrients vary widely in different parts of the sea and at different times of the year. All animals and plants are represented except amphibians, centipedes, millipedes, and insects. Life may have originated in the intertidal zone of the marine biome, which is the zone between the high and low tide.

The marine biome is made up of four zones:

1) The intertidal zone supports a variety of organisms because of the high and low tide.

2) The littoral zone is beyond the intertidal zone. It includes many species of aquatic organisms especially producers.

3) The open sea zone – The upper layer of this zone supports a tremendous amount of producers and therefore many consumers. The lower layer, though, supports only a few scavengers and their predators.

4) The ocean floor contains bacteria of decay and worms.

G) Freshwater Zones – Freshwater zones are divided into standing water-lakes, ponds and swamps, and running water, rivers, creeks, and springs. Freshwater zones are characterized by an assortment of animals and plants. Aquatic life is most prolific in the littoral zones of lakes. Freshwater zones change much more rapidly than other biomes.

Biogeochemical Cycles

Biogeochemical cycles involve the movement of mineral ions and molecules in and out of ecosystems. Most ions enter the living realm at the producer level.

The Nitrogen Cycle – Outside of life, nitrogen occurs in exchange pools and reservoirs. The largest reservoir is N_2 in the atmosphere, but is only available to nitrogen–fixers. Nitrate ions are made available in soil exchange pools.

Phosphorus and Calcium Cycles – Cycles of calcium and phosphorus occur between living organisms and water. The two ions are taken up in soluble phosphate and calcium ions. Phosphates are used in producing ATP, nucleic acids, phospholipids, and tooth and shell materials, while calcium is essential to bone and shell development and in membrane activity.

The Carbon Cycle – Carbon is an essential part of nearly all the molecules of life. The principal exchange pool on land consists of carbon dioxide gas while the source in the water is dissolved carbon dioxide gas and carbonate ion. A large reservoir occurs in the form of limestone and fossil fuels.

E. Heredity

Meiosis

Meiosis – Meiosis consists of two successive cell divisions with only one duplication of chromosomes. This results in daughter cells with a haploid number of chromosomes or one-half of the chromosome number in the original cell. This process occurs during the formation of gametes and in spore formation in plants.

A) Spermatogenesis – This process results in sperm cell formation with four immature sperm cells with a haploid number of chromosomes.

B) Oogenesis – This process results in egg cell formation with only one immature egg cell with a haploid number of chromosomes, which becomes mature and larger as yolk forms within the cell.

First Meiotic Division –

A) Interphase I – Chromosome duplication begins to occur during this phase.

B) Prophase I – During this phase, the chromosomes shorten and thicken and synapsis occurs with pairing of homologous chromosomes. Crossing-over between non-sister chromatids will also occur. The centrioles will migrate to opposite poles and the nucleolus and nuclear membrane begin to dissolve.

C) Metaphase I – The tetrads, composed of two doubled homologous chromosomes, migrate to the equatorial plane during metaphase I.

D) Anaphase I – During this stage, the paired homologous chromosomes separate and move to opposite poles of the cell. Thus, the number of chromosome types in each resulting cell is reduced to the haploid number.

E) Telophase I – Cytoplasmic division occurs during telophase I. The formation of two new nuclei with half the chromosomes of the original cell occurs.

F) Prophase II – The centrioles that had migrated to each pole of the parental cell, now incorporated in each haploid daughter cell, divide, and a new spindle forms in each cell. The chromosomes move to the equator.

G) Metaphase II – The chromosomes are lined up at the equator of the new spindle, which is at a right angle to the old spindle.

H) Anaphase II – The centromeres divide and the daughter chromatids, now chromosomes, separate and move to opposite poles.

I) Telophase II – Cytoplasmic division occurs. The chromosomes gradually return to the dispersed form and a nuclear membrane forms.

Mendelian Genetics and Laws, Probability

By studying one single trait at a time in garden peas, Gregor Mendel, in 1857, was able to discover the basic laws of genetics.

A) Definitions

1) A gene is the part of a chromosome that codes for a certain hereditary trait.

2) A chromosome is a filamentous or rod-shaped body in the cell nucleus that contains the genes.

3) A genotype is the genetic makeup of an organism, or the set of genes that it possesses.

4) A phenotype is the outward, visible expression of the hereditary makeup of an organism.

5) Homologous chromosomes are chromosomes bearing genes for the same characters.

6) A homozygote is an organism possessing an identical pair of alleles on homologous chromosomes for a given character or for all given characters.

7) A heterozygote is an organism possessing different alleles on homologous chromosomes for a given character or for all given characters.

8) Crossing over means that paired chromosomes may break and their fragments reunite in new combinations.

9) Translocations are the shifting of gene positions in chromosomes that may result in a change in the serial arrangement of genes. In general, it is the transfer of a chromosome fragment to a non-homologous chromosome.

10) Linkage is the tendency of two or more genes on the same chromosome to cause the traits they control to be inherited together.

An Abstract of the Data Obtained by Mendel
from His Breeding Experiments with Garden Peas

Parental Characters	First Generation	Second Generation	Ratios
Yellow seeds × green seeds	all yellow	6022 yellow:2001 green	3.01:1
Round seeds × wrinkled seeds	all round	5474 round:1850 wrinkled	2.96:1
Green pods × yellow pods	all green	428 green:152 yellow	2.82:1
Long stems × short stems	all long	787 long:277 short	2.84:1
Axial flowers × terminal flowers	all axial	651 axial:207 terminal	3.14:1
Inflated pods × constricted pods	all inflated	882 inflated:299 constricted	2.95:1
Red flowers × white flowers	all red	705 red:224 white	3.15:1

The 3:1 ratio resulting from this data enabled Mendel to recognize that the offspring of each plant had two factors for any given characteristic instead of a single factor.

11) Alleles are types of alternative genes that occupy a given locus on a chromosome; they are matching genes that can control contrasting characters.

12) Genetic mutation is a change in an allele or segment of a chromosome that may give rise to an altered genotype, which often leads to the expression of an altered phenotype.

13) A codon is a sequence of three adjacent nucleotides that codes for a single amino acid.

B) Laws of Genetics

1) Law of Dominance – Of two contrasting characteristics, the dominant one may completely mask the appearance of the recessive one.

2) Law of Segregation and Recombination – Each trait is transmitted as an unchanging unit, independent of other traits, thereby giving the recessive

traits a chance to recombine and show their presence in some of the offspring.

3) Law of Independent Assortment – Each character for a trait operates as a unit and the distribution of one pair of factors is independent of another pair of factors linked on different chromosomes.

Patterns of Inheritance, Chromosomes, Genes, and Alleles

In 1900, Walter Sutton compared the behavior of chromosomes with the behavior of the hereditary characters that Mendel had proposed and formulated the chromosome principle of inheritance.

The Chromosome Principle of Inheritance:

A) Chromosomes and Mendelian factors exist in pairs.

B) The segregation of Mendelian factors corresponds to the separation of homologous chromosomes during the reduction division stage of meiosis.

C) The recombination of Mendelian factors corresponds to the restoration of the diploid number of chromosomes at fertilization.

D) The factors that Mendel described as passing from parent to offspring correspond to the passing of chromosomes into gametes which then unite and develop into offspring.

E) The Mendelian idea that two sets of characters present in a parent assort independently corresponds to the random separation of the two sets of chromosomes as they enter a different gamete during meiosis.

Sutton's chromosome principle of inheritance states that the hereditary characters, or factors, that control heredity are located in the chromosomes. By 1910, the factors of heredity were called genes.

CHAPTER 3

ORGANISMAL AND POPULATION BIOLOGY

A. THE INTERRELATIONSHIP OF LIVING THINGS

Taxonomy and Systematics

Taxonomy is the science of classification, and systematics is the science of evolutionary relationships.

The Five Kingdom Classification Scheme

The largest taxon is the kingdom. In the scheme of classification, there are five kingdoms: the Prokaryota, Protista, Fungi, Plantae, and Animalia. They are subdivided into descending levels of taxa: phylum, class, order, family, genus, and subspecies (sometimes varieties and races).

B. DIVERSITY AND CHARACTERISTICS OF THE KINGDOMS MONERA

Prokaryota

Bacteria and blue-green algae are included in the kingdom Monera. These cells having no nuclear membrane and only a single chromosome are termed prokaryotes. Both blue-green algae and bacteria lack membrane bounded subcellular organelles such as mitochondria and chloroplasts.

Protista

Most protists are unicellular. Some are composed of colonies. All protists are eukaryotes. That is, they are characterized by nuclei bounded by a nuclear membrane. Examples of protists are flagellates, the protozoans, and slime molds.

Algae—photosynthetic protists—range from single-celled to complex multicellular forms. Most are aquatic, living in all of the earth's waters, and in a few instances on land. Included are the vast, floating, microscopic marine phytoplankton and the marine seaweeds and kelps.

Fungi

Fungi are multicellular, nonmotile heterotrophs, lacking tissue organization except in their reproductive structures. Some are coenocytic (made up of a multinucleated mass of cytoplasm without subdivision into cells), while others are cellular. Most have chitinous cell walls, while a few have walls of cellulose.

C. THE KINGDOM PLANTAE

Diversity, Classification, Phylogeny

Plants are characterized as being multicellular and photosynthetic, most with tissue, organ, and system organization (roots, stems, leaves, flowers). They develop from protected embryos, contain chlorophylls a and b, produce starches, and have walls of cellulose.

Plants are classified into five categories:

A) Nonvascular Plants – Characterized as having little or no organized tissue for conducting water and food (xylem and phloem), e.g. mosses.

B) Seedless Vascular Plants – Characterized as having organized vascular tissue, limited root development, many primitive traits, chlorophyll a and b, e.g. ground pine, horsetails, and ferns.

C) Seed-Producing Vascular Plants – Characterized as having seeds, organized vascular tissues, and extensive root, stem and leaf development.

D) Gymnosperms – Characterized as having seeds without surrounding fruit, tracheids only for water conduction, terrestrial, e.g. maidenhair tree, Cycads, Gnetum, pines, redwoods, firs, junipers, larch, cypress, and hemlock.

E) Angiosperms – Characterized as having flowers, seeds with surrounding fruit, e.g. magnolia, cabbage, tobacco, cotton, iris, orchids, and grains.

Adaptations to Land

To accommodate to land, plants (except for the nonvascular plants) grew much larger due to the presence of vascular tissue, including the woody, hardened xylem, and tough accompanying supporting tissues. Apparently large size provided a novel way of adapting to the terrestrial environment. Also, deep root systems allowed the plants to obtain water and allowed them to grow larger.

The Life Cycle (Life History): Alternation of Generations in Moss, Fern, Pine, Angiosperms (Flowering Plants)

Alternation of generations in plants is characterized by the existence of two phases in the life of a single individual, including a diploid spore-producing phase and a haploid gamete-producing phase.

Plant	Alternation of Generation
moss	separate generations, gametophyte usually dominant, swimming sperm, egg, and zygote protected.
fern	separate generations, dominant sporophyte, minute gametophyte, swimming sperm.
pine	nonmotile sperm, wind-pollinated.
angiosperms	nonmotile sperm, both wind- and insect-pollinated.

Anatomy, Morphology, and Physiology of Vascular Plants

A mature vascular plant possesses several distinct cell types which group together in tissues. The major plant tissues include epidermal, parenchyma, sclerenchyma, chlorenchyma, vascular, and meristematic.

Summary of Plant Tissues

Tissue	Location	Functions
Epidermal	Root	Protection-Increases absorption area
	Stem	Protection-Reduces H_2O loss
	Leaf	Protection-Reduces H_2O loss Regulates gas exchange
Parenchyma	Root, stem, leaf	Storage of food and H_2O
Sclerenchyma	Stem and leaf	Support
Chlorenchyma	Leaf and young stems	Photosynthesis
Vascular a. Xylem	Root, stem, leaf	Upward transport of fluid
b. Phloem	Leaf, root stem	Downward transport of fluid
Meristematic	Root and stem	Growth; formation of xylem, phloem, and other tissues

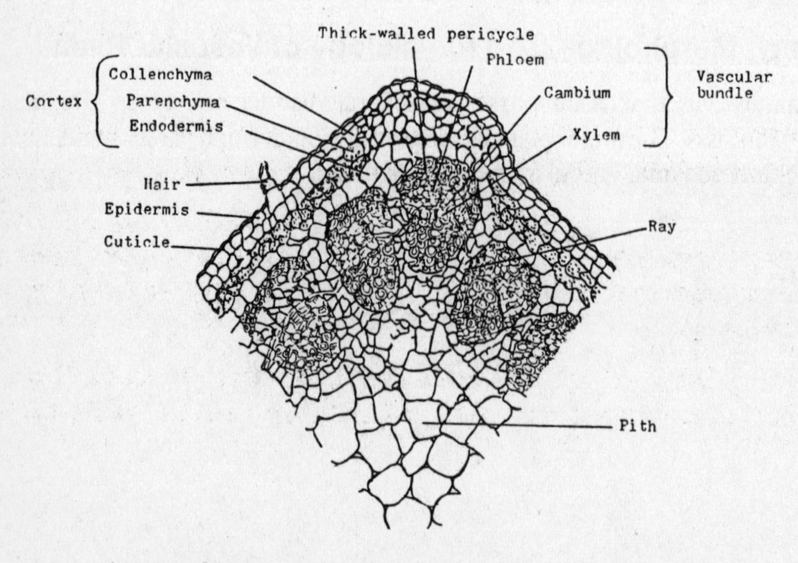

A sector of a cross section of a stem from a herbaceous dicot, alfalfa.

Reproduction and Growth in Seed Plants: Seed Formation, Seed Structure, Germination, Growth

Mitosis and differentiation in the primary endosperm and zygote will produce the embryo, which along with the food supply and seed coat, make up the seed.

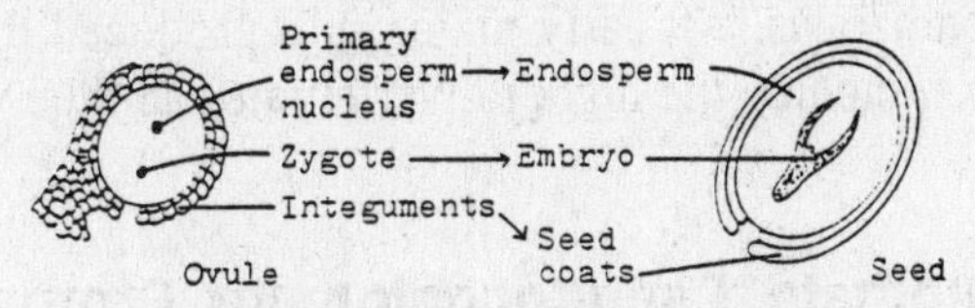

Development of an ovule into a seed.

For many seeds, germination (the emergence from dormancy) simply requires the uptake of water, suitable temperatures, and the availability of oxygen, but others require special conditions, such as exposure to freezing temperatures, fire, abrasion, or exposure to animal digestive enzymes. The growth of the plant is dependent on three key hormones.

Plant Hormones: Types, Functions, Effects on Plant Growth

Auxins – These plant growth regulators stimulate the elongation of specific plant cells and inhibit the growth of other plant cells.

Gibberellins – In some plants, gibberellins are involved in the stimulation of flower formation. They also increase the stem length of some plant species and the size of fruits. Gibberellins also stimulate the germination of seeds.

Cytokinins – Cytokinins increase the rate of cell division and stimulate the growth of cells in a tissue culture. They also influence the shedding of leaves and fruits, seed germination, and the pattern of branch growth.

Environmental Influences on Plants and Plant Responses to Stimuli: Tropisms, Photoperiodism

Auxin is involved in tropisms, or growth responses. They involve positive phototropism (bending or growing toward light), negative phototropism (bending or growing away from light), geotropism (influenced by gravity), and thigmotropism (mechanical or touch).

Photoperiodism is any response to changing lengths of night or day. Flowering in plants is often photoperiodic, but so far no specific hormonal mechanism has been found.

D. THE KINGDOM ANIMALIA

Diversity, Classification, Phylogeny

The phylogenetic tree representing animal evolution reveals the separate origins of metazoans and parazoans. An early major split produced the protostomes and deuterostomes, which include the higher invertebrates and the vertebrates, respectively.

Survey of Acoelomate, Pseudocoelomate, Protostome, and Deuterostome Phyla

The acoelomates are animals that have no coelom (body cavity.) They include the phylum Platyhelminthes (flatworms) and the phylum Nematoda. In acoelomate animals, the space between the body wall and the digestive tract is not a cavity, as in higher animals, but is filled with muscle fibers and a loose tissue of mesenchymal origin called parenchyma, both derived from the mesoderm.

The pseudocoelomates consist of the following phyla: Nematoda, Rotifera, Gastrotricha, Nematomorpha, and Acanthocephala. These animals have a body cavity which is not entirely lined with peritoneum.

A major division in animal evolution produced the protostomes and deuterostomes. Each has bilateral symmetry, a one way gut, and a true coelom (eucoelomate). During primitive gut formation in protostome embryos, the blastopore forms the mouth of the animal, while in deuterostomes this is the anal area.

Structure and Function of Tissues, Organs, and Systems Especially in Vertebrates

The cells that make up multicellular organisms become differentiated in many ways. One or more types of differentiated cells are organized into tissues. The basic tissues of a complex animal are the epithelial, connective, nerve, muscle, and blood tissues.

Summary of Animal Tissues

Tissue	Location	Functions
Epithelial	Covering of body Lining internal organs	Protection Secretion
Muscle		
Skeletal	Attached to skeleton bones	Voluntary movement
Smooth	Walls of internal organs	Involuntary movement
Cardiac	Walls of heart	Pumping blood
Connective		
Binding	Covering organs, in tendons and ligaments	Holding tissues and organs together
Bone	Skeleton	Support, protection, movement
Adipose	Beneath skin and around internal organs	Fat storage, insulation, cushion
Cartilage	Ends of bone, part of nose and ears	Reduction of friction, support
Nerve	Brain	Interpretation of impulses, mental activity
	Spinal cord, nerves, ganglions	Carrying impulses to and from all organs
Blood	Blood vessels, heart	Carrying materials to and from cells, carrying oxygen, fighting germs, clotting

THE HUMAN DIGESTIVE SYSTEM

The digestive system of man consists of the alimentary canal and several glands. This alimentary canal consists of the oral cavity (mouth), pharynx, esophagus, stomach, small intestine, large intestine, and the rectum.

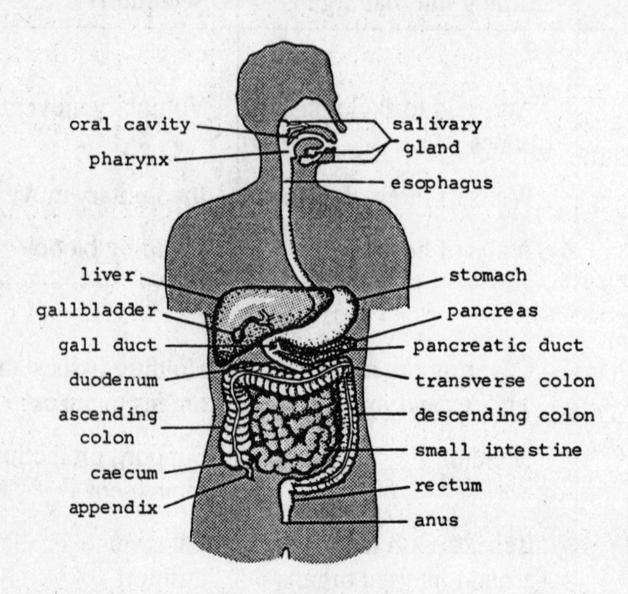

Human digestive system. (The organs are slightly displaced, and the small intestine is greatly shortened.)

A) Oral Cavity (mouth) – The mouth cavity is supported by jaws and is bound on the sides by the teeth, gums, and cheeks. The tongue binds the bottom and the palate binds the top. Food is pushed between the teeth by the action of the tongue so it can be chewed and swallowed. Saliva is the digestive juice secreted that begins the chemical phase of digestion.

B) Pharynx – Food passes from the mouth cavity into the pharynx where the digestive and respiratory passages cross. Once food passes the upper part of the pharynx, swallowing becomes involuntary.

C) Esophagus – Whenever food reaches the lower part of the pharynx, it enters the esophagus and peristalsis pushes the food further down the esophagus into the stomach.

D) Stomach – The stomach has two muscular valves at both ends: the cardiac sphincter which controls the passage of food from the esophagus into the stomach, and the pyloric sphincter which is responsible for the control of the passage of partially digested food from the stomach to the small intestine. Gastric juice is also secreted by the gastric glands lining the stomach walls. Gastric juice begins the digestion of proteins.

E) Pancreas – The pancreas is the gland formed by the duodenum and the under surface of the stomach. It is responsible for producing pancreatic fluid which aids in digestion. Sodium bicarbonate, an amylase, a lipase, trypsin, chymotrypsin, carboxypeptidase, and nucleases are all found in the pancreatic fluid.

F) Small Intestine – The small intestine is a narrow tube between 20 and 25 feet long divided into three sections: the duodenum, the jejunum, and the ileum. The final digestion and absorption of disaccharides, peptides, fatty acids, and monoglycerides is the work of villi, small finger-line projections, which line the small intestine.

G) Liver – Even though the liver is not an organ of digestion, it does secrete bile which aids in digestion by neutralizing the acid chyme from the stomach and emulsifying fats. The liver is also responsible for the chemical destruction of excess amino acids, the storage of glycogen, and the breakdown of old red blood cells.

H) Large Intestine – The large intestine receives the liquid material that remains after digestion and absorption in the small intestine have been completed. However, the primary function of the large intestine is the reabsorption of water.

Ingestion and Digestion in Other Organisms

Hydra – The hydra possesses tentacles which have stinging cells (nematocysts) which shoot out a poison to paralyze the prey. If successful in capturing an animal, the tentacles push it into the hydra's mouth. From there, the food enters the gastric cavity. The hydra uses both intracellular and extracellular digestion.

Earthworm – As the earthworm moves through soil, the suction action of the pharynx draws material into the mouth cavity. Then from the mouth, food goes into the pharynx, the esophagus, and then the crop which is a temporary storage area. This food then passes into a muscular gizzard where is is ground and churned. The food mass finally passes into the intestine; any undigested material is eliminated through the anus.

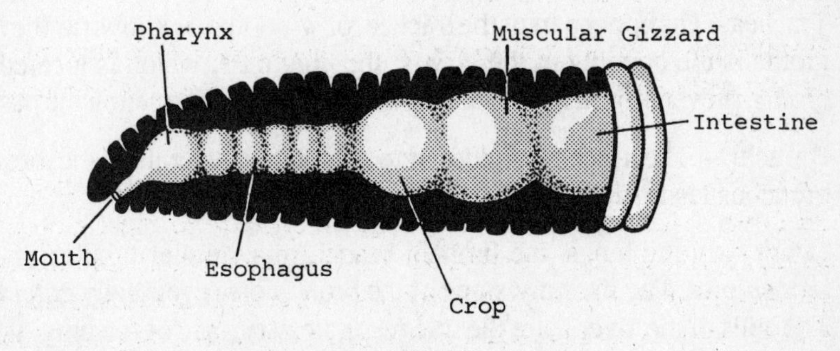

The digestive system of the earthworm.

Grasshopper – The grasshopper is capable of consuming large amounts of plant leaves. This plant material must first pass through the esophagus into the crop, a temporary storage organ. It then travels to the muscular gizzard where food is ground. Digestion takes place in the stomach. Enzymes secreted by six gastric glands are responsible for digestion. Absorption takes place mainly in the stomach. Undigested material passes into the intestine, collects in the rectum, and is eliminated through the anus.

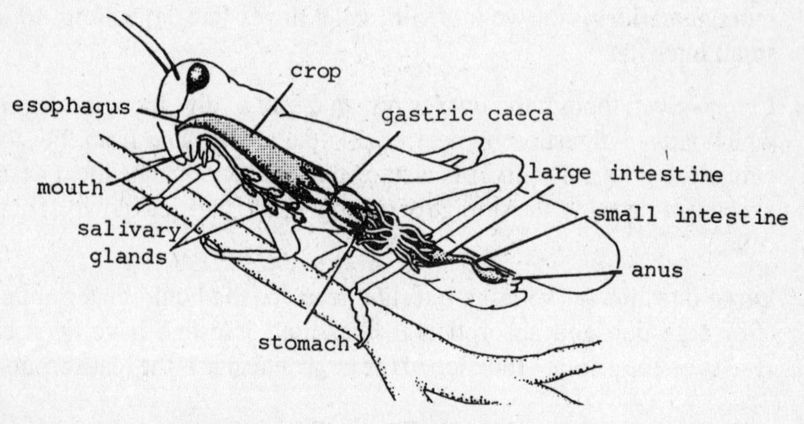

The digestive system of the grasshopper.

RESPIRATION IN HUMANS

The respiratory system in humans begins as a passageway in the nose. Inhaled air then passes through the pharynx, the trachea, the bronchi, and the lungs.

A) Nose – The nose is better adapted to inhale air than the mouth. The nostrils, the two openings in the nose, lead into the nasal passages which are lined by the mucous membrane. Just beneath the mucous membrane are capillaries which warm the air before it reaches the lungs.

B) Pharynx – Air passes via the nasal cavities to the pharynx where the paths of the digestive and respiratory systems cross.

C) Trachea – The upper part of the trachea, or windpipe, is known as the larynx. The glottis is the opening in the larynx; the epiglottis, which is located above the glottis, prevents food from entering the glottis and obstructing the passage of air.

D) Bronchi – The trachea divides into two branches called the bronchi. Each bronchus leads into a lung.

E) Lungs – In the lungs, the bronchi branch into smaller tubules known as the bronchioles. The finer divisions of the bronchioles eventually enter the alveoli. The cells of the alveoli are the true respiratory surface of the lung. It is here that gas exchange takes place.

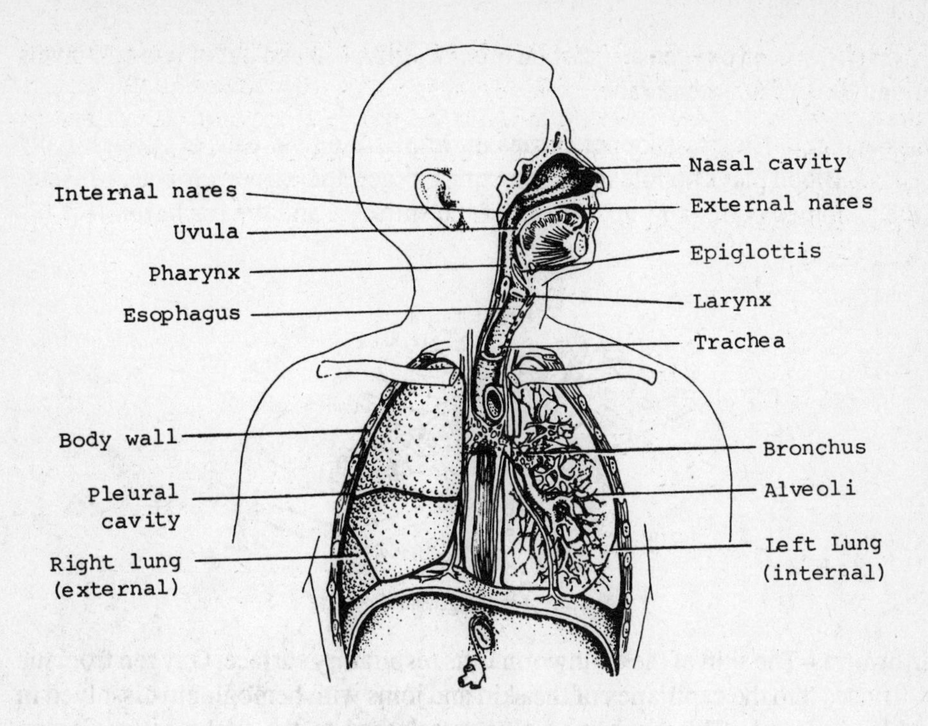

The human respiratory system.

RESPIRATION IN OTHER ORGANISMS

Protozoa

A) Amoeba – Simple diffusion of gases between the cell and water is sufficient to take care of the respiratory needs of the amoeba.

B) Paramecium – The paramecium takes in dissolved oxygen and releases dissolved carbon dioxide directly through the plasma membrane.

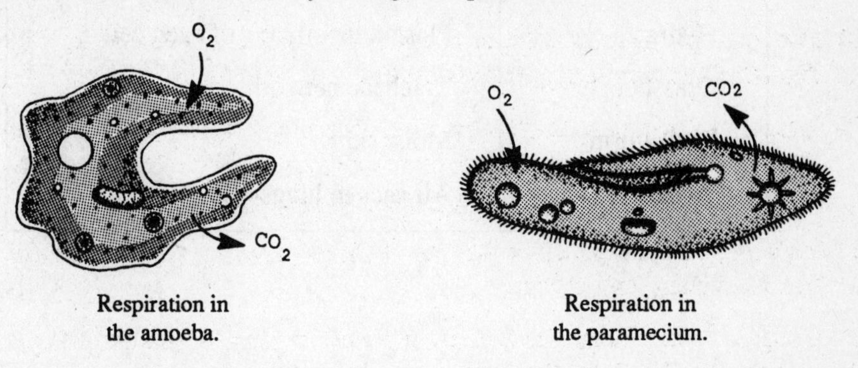

Respiration in the amoeba.

Respiration in the paramecium.

Hydra – Dissolved oxygen and carbon dioxide diffuse in and out of two cell layers through the plasma membrane.

Grasshopper – The grasshopper carries on respiration by means of spiracles and tracheae. Blood plays no role in transporting oxygen and carbon dioxide. Muscles of the abdomen pump air into and out of the spiracles and the tracheae.

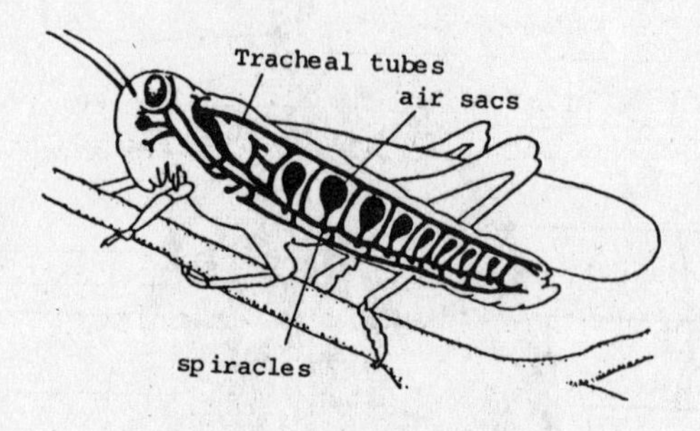

Respiration in the grasshopper.

Earthworm – The skin of the earthworm is its respiratory surface. Oxygen from the air diffuses into the capillaries of the skin and joins with hemoglobin dissolved in the blood plasma. This oxyhemoglobin is released to the tissue cells. Carbon dioxide from the tissue cells diffuses into the blood. When the blood reaches the capillaries in the skin again, the carbon dioxide diffuses through the skin into the air.

Comparison of Various Respiratory Surfaces Among Organisms

Organism	Respiratory Surface Present
Protozoan	Plasma membrane
Hydra	Plasma membrane of each cell
Grasshopper	Tracheae network
Earthworm	Moist skin
Human	Air sacs in lungs

THE HUMAN CIRCULATORY SYSTEM

Humans have a closed circulatory system in which the blood moves entirely within the blood vessels. This circulatory system consists of the heart, blood, veins, arteries, capillaries, lymph, and lymph vessels.

The heart is a pump covered by a protective membrane known as the pericardium and divided into four chambers. These chambers are the left and right atrium, and the left and right ventricles.

Atria – The atria are the upper chambers of the heart that receive blood from the superior and inferior vena cava. This blood is then pumped to the lower chambers or ventricles.

Ventricles – The ventricles have thick walls as compared to the thin walls of the atria. They must pump blood out of the heart to the lungs and other distant parts of the body.

The heart also contains many important valves. The tricuspid valve is located between the right atrium and the right ventricle. It prevents the backflow of blood into the atrium after the contraction of the right ventricle. The bicuspid valve, or mitral valve, is situated between the left atrium and the left ventricle. It prevents the backflow of blood into the left atrium after the left ventricle contracts.

TRANSPORT MECHANISMS IN OTHER ORGANISMS

Protozoans – Most protozoans are continually bathed by food and oxygen because they live in water or another type of fluid. With the process of cyclosis or diffusion, digested materials and oxygen are distributed within the cell, and water and carbon dioxide are removed. Proteins are transported by the endoplasmic reticulum.

Hydra – Like the protozoans, materials in the hydra are distributed to the necessary organelles by diffusion, cyclosis, and by the endoplasmic reticulum.

Earthworm – The circulatory system of the earthworm is known as a "closed" system because the blood is confined to the blood vessels at all times. A pump that forces blood to the capillaries consists of five pairs of aortic loops. Contraction of these loops forces blood into the ventral blood vessel. This ventral blood vessel transports blood toward the rear of the worm. The dorsal blood vessel forces blood back to the aortic loops at the anterior end of the worm.

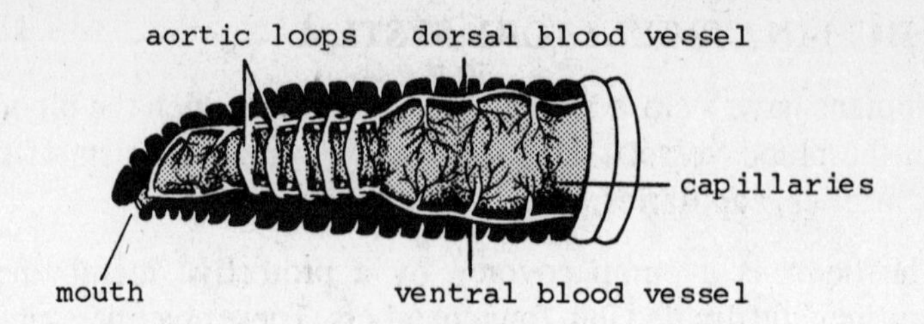

"Closed" circulatory system of the earthworm.

THE HUMAN ENDOCRINE SYSTEM

The major glands of the human endocrine system include the thyroid gland, parathyroid glands, pituitary gland, pancreas, adrenal glands, pineal gland, thymus gland, and the sex glands.

A) Thyroid Gland – The thyroid gland is a two-lobed structure located in the neck. It is responsible for the secretion of the hormone thyroxin. Thyroxin increases the rate of cellular oxidation and influences growth and development of the body.

B) Parathyroid Glands – The parathyroid glands are located in back of the thyroid gland. They secrete the hormone parathormone which is responsible for regulating the amount of calcium and phosphate salts in the blood.

C) Pituitary Gland – The pituitary gland is located at the base of the brain. It consists of three lobes: the anterior lobe; the intermediate lobe, which is only a vestige in adulthood; and the posterior lobe.

   1) Hormones of the Anterior Lobe

      a) growth hormone – stimulates growth of bones.

      b) thyroid-stimulating hormone (TSH) – stimulates the thyroid gland to produce thyroxin.

      c) prolactin – regulates development of the mammary glands of a pregnant female and stimulates secretion of milk in a woman after childbirth.

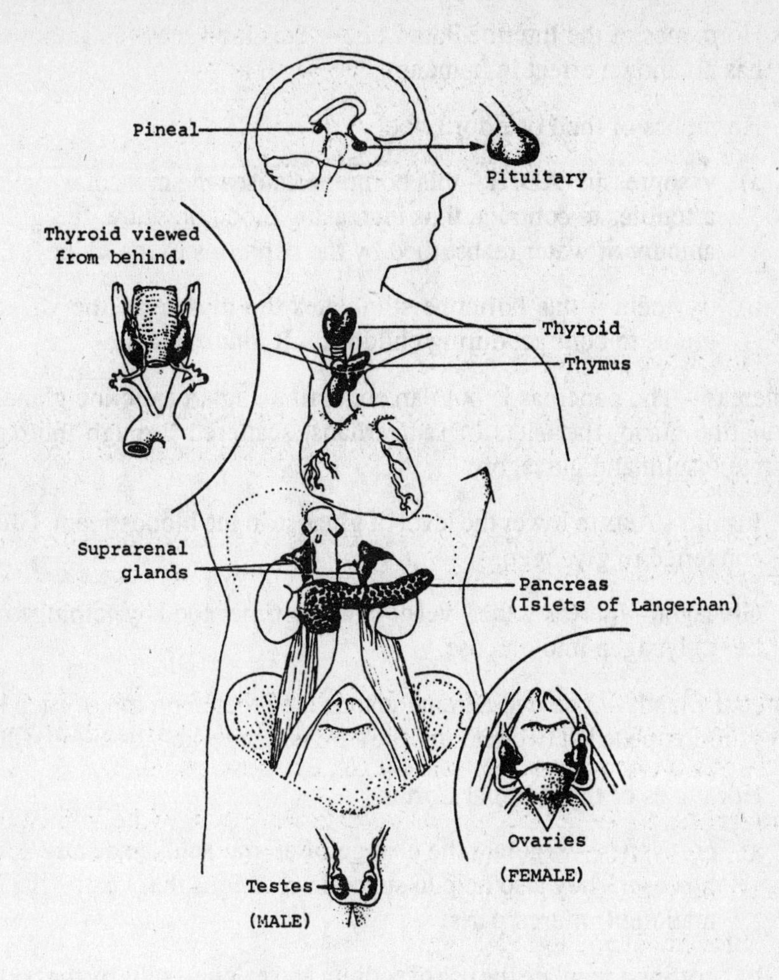

The human endocrine system.

d) adrenocorticotrophic hormone (ACTH) – stimulates the secretion of hormones by the cortex of the adrenal glands.

e) follicle-stimulating hormone (FSH) – this hormone acts upon the gonads, or sex organs.

f) luteinizing hormone (LH) – in the male, LH causes the cells in the testes to secrete androgens. In females, LH causes the follicle in an ovary to change into the corpus luteum.

2) Hormones of the Intermediate Lobe – This lobe secretes a hormone that has no known effect in humans.

3) Hormones of the Posterior Lobe

a) vasopressin (ADH) – this hormone causes the muscular walls of the arterioles to contract, thus increasing blood pressure. It regulates the amount of water reabsorbed by the nephrons in the kidney.

b) oxytocin – this hormone stimulates the muscle of the walls of the uterus to contract during childbirth. It induces labor.

D) Pancreas – The pancreas is both an endocrine and an exocrine gland. As an endocrine gland, the islets of Langerhans, scattered through the pancreas, secrete insulin and glucagon.

1) Insulin – Acts to lower the level of glucose in the bloodstream. Glucose is converted to glycogen.

2) Glucagon – Increases the level of glucose in the blood by helping to change liver glycogen into glucose.

E) Adrenal Glands – The two adrenal glands are located on top of each kidney. They are composed of two regions: the adrenal cortex and the adrenal medulla.

1) Hormones of the Adrenal Cortex

a) cortisones – regulate the change of amino acids and fatty acids into glucose. They also help to suppress reactions that lead to the inflammation of injured parts.

b) cortins – regulate the use of sodium and calcium salts by the body cells.

c) sex hormones – they are similar in chemical composition to hormones secreted by sex glands.

2) Hormones of the Adrenal Medulla

a) epinephrine – this hormone is responsible for the release of glucose from the liver, the relaxation of the smooth muscles of the bronchioles, dilation of the pupils of the eye, a reduction in the clotting time of blood, and an increase in the heartbeat rate, blood pressure, and respiration rate.

b) norepinephrine – this hormone is responsible for the constriction of blood vessels.

F) Pineal Gland – The pineal gland is attached to the brain above the cerebellum. It is responsible for the production of melatonin whose role in humans is uncertain.

G) Thymus Gland – The thymus gland is located under the breastbone. Although there is no convincing evidence for its role in the human adult, it does secrete thymus hormone in infants which stimulates the formation of an antibody system.

H) Sex Glands – These glands include the testes of the male and the ovaries of the female.

   1) Testes – Luteinizing hormone stimulates specific cells of the testes to secrete androgens. Testosterone, which controls the development of male secondary sex characteristics, is the principal androgen.

   2) Ovaries – Estrogen is secreted from the cells which line the ovarian follicle. This hormone is responsible for the development of female secondary sex characteristics.

Human Endocrine Glands and Their Functions

Gland	Hormone	Function
Pituitary Anterior lobe	Growth hormone	Stimulates growth of skeleton
	FSH	Stimulates follicle formation in ovaries and sperm formation in testes
	LH	Stimulates formation of corpus luteum in ovaries and secretion of testosterone in testes
	TSH	Stimulates secretion of thyroxin from thyroid gland
	ACTH	Stimulates secretion of cortisone and cortin from adrenal cortex

Human Endocrine Glands and Their Functions (Cont'd)

Gland	Hormone	Function
Pituitary Anterior lobe (cont'd)	Prolactin	Stimulates secretion of milk in mammary glands
Posterior lobe	Vasopressin (ADH)	Controls narrowing of arteries and rate of water absorption in kidney tubules
	Oxytocin	Stimulates contraction of smooth muscle of uterus
Thyroid	Thyroxin	Controls rate of metabolism and physical and mental development
	Calcitonin	Controls calcium metabolism
Parathyroids	Parathormone	Regulates calcium and phosphate level of blood
Islets of Langerhans Beta cells	Insulin	Promotes storage and oxidation of glucose
Alpha cells	Glucagon	Releases glucose into bloodstream
Thymus	Thymus hormone	Stimulates formation of antibody system
Adrenal Cortex	Cortisones	Promote glucose formation from amino acids and fatty acids
	Cortins	Control water and salt balance

Human Endocrine Glands and Their Functions (Cont'd)

Gland	Hormone	Function
Adrenal Cortex (cont'd)	Sex hormones	Influence sexual development
Medulla	Epinephrine (adrenalin) or norepinephrine (noradrenalin)	Releases glucose into bloodstream, increases rate of heartbeat, increases rate of respiration, reduces clotting time, relaxes smooth muscle in air passages
Gonads Ovaries, follicle cells	Estrogen	Controls female secondary sex characteristics
Corpus luteum cells	Progesterone	Helps maintain attachment of embryo to mother
Testes	Testosterone	Controls male secondary sex characteristics

NERVOUS SYSTEM

The nervous system is a system of conduction that transmits information from receptors to appropriate structures for action.

Neurons – The unit of structure that conducts electrochemical impulses over a certain distance. In many neurons, the nerve impulses are generated in the dendrites. These impulses are then conducted along the axon, which is a long fiber. A myelin sheath covers the axon.

A) Sensory Neurons – Sensory neurons conduct impulses from receptors to the central nervous system.

B) Interneurons – Interneurons are always found within the spinal cord and the brain. They form the intermediate link in the nervous system pathway.

C) Motor Neurons – Motor neurons conduct impulses from the central nervous system to the effectors which are muscles and glands. They will bring about the responses to the stimulus.

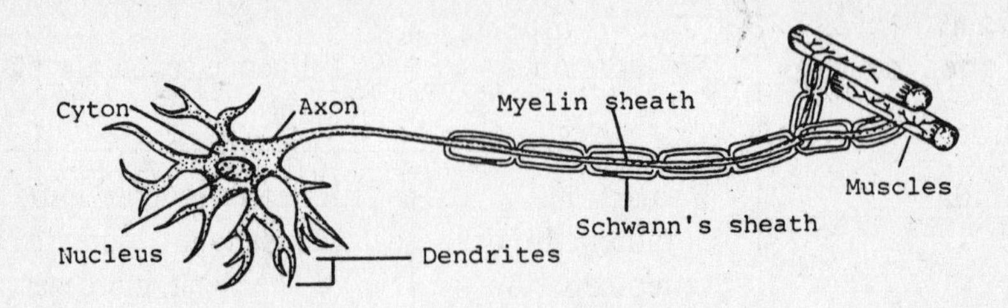

The structure of a neuron.

Nerve Impulse – The signal that is transmitted from one neuron to another. When a neuron is not stimulated, the outside of the neuron is positively charged and the inside is negatively charged. However, when there is neuron stimulation, the inside of the neuron is temporarily positively charged and the outside is temporarily negatively charged. This marks the beginning of the generation and flow of the nerve impulse.

Synapse – The junction between the axon of one neuron and the dendrite of the next neuron in line. An impulse is transmitted across the synaptic gap by a specific chemical transmitter which is acetylcholine. When a nerve impulse reaches the end brush of the first axon, the end brush secretes acetylcholine into the synaptic gap. It is here that the acetylcholine changes the permeability of the dendrite's membrane of the second neuron. As soon as the acetylcholine is no longer needed, it is decomposed by the enzyme acetylcholinesterase.

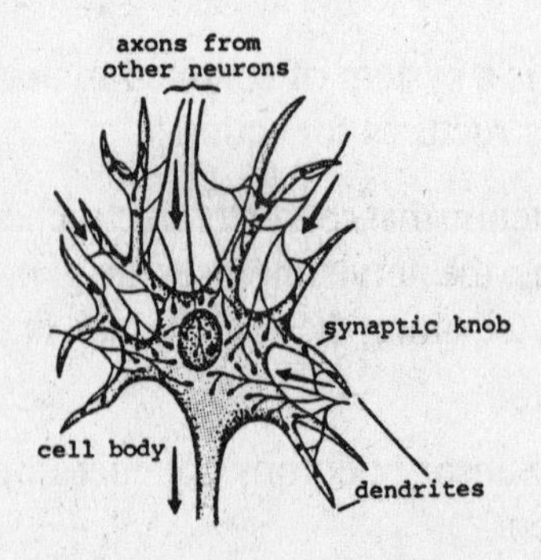

Nerve impulse across a synapse.

Reflex Arc – The unit of function of the nervous system. It is formed by a sequence of sensory neurons, interneurons, and motor neurons which conduct the nerve impulses for the given reflex.

The Human Nervous System

A) The Central Nervous System – The brain and the spinal cord comprise the central nervous system. The brain is divided into three regions: the forebrain, the midbrain, and the hindbrain. Each of these regions has a specific function attributed to the particular lobe.

1) Brain

a) Forebrain – The cerebrum is the most prominent part of the forebrain and is divided into two hemispheres. It also has four major areas known as the sensory area, motor area, speech area, and association area. The thalamus, hypothalamus, pineal gland, and part of the pituitary gland are also part of the human forebrain.

b) Midbrain – The midbrain is one of the smallest regions of the human brain. Its function is to relay nerve impulses between the two other brain regions: the forebrain and the hindbrain. It also aids in the maintenance of balance.

c) Hindbrain – The medulla oblongata and the cerebellum are the two main regions of the hindbrain.

1) medulla oblongata – controls reflex centers for respiration and heartbeat, coughing, swallowing, and sneezing.

2) cerebellum – coordinates locomotor activity in the body initiated by impulses originating in the forebrain.

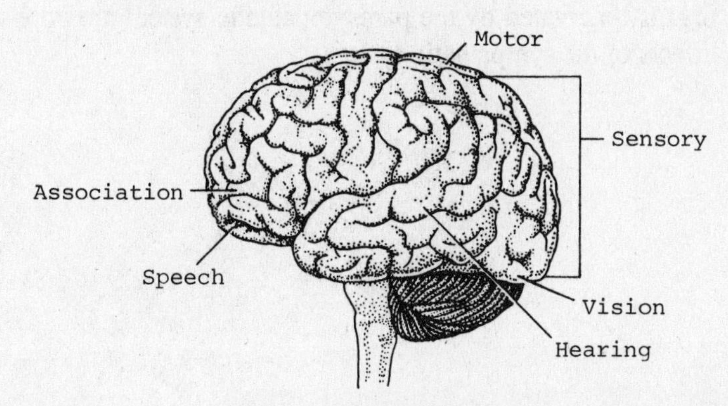

The major areas of the cerebrum.

2) Spinal Cord – The spinal cord runs from the medulla down through the backbone. Throughout its length, it is enclosed by three meninges and by the spinal column vertebrae. Running vertically in the spinal cord center is a narrow canal filled with cerebrospinal fluid.

   The spinal cord contains control centers for reflex acts below the neck, and it provides the major pathway for impulses between the peripheral nervous system and the brain. It is also a connecting center between sensory and motor neurons.

B) The Peripheral Nervous System – The peripheral nervous system is composed of nerve fibers which connect the brain and the spinal cord (central nervous system) to the sense organs, glands, and muscles. It can be subdivided into the somatic nervous system and the autonomic nervous system.

1) Somatic Nervous System – The somatic nervous system consists of nerves which transmit impulses from receptors to the central nervous system and from the central nervous system to the skeletal muscles of the body.

2) Autonomic Nervous System – The autonomic nervous system is composed of sensory and motor neurons which run between the central nervous system and various internal organs, such as the heart, glands, and intestines. It regulates internal responses which keep the internal environment constant. It is subdivided into two smaller systems.

   a) sympathetic system – a branch of the autonomic nervous system with motor neurons arising from the spinal cord. This system accelerates heartbeat rate, constricts arteries, slows peristalsis, relaxes the bladder, dilates breathing passages, dilates the pupil, and increases secretion.

   b) parasympathetic system – a branch of the autonomic nervous system which consists of fibers arising from the brain. The effectors on the organs innervated by the parasympathetic system are opposite to the effects of the sympathetic system.

Actions of the Autonomic Nervous System

Organ Innervated	Sympathetic Action	Parasympathetic Action
Heart	Accelerates heartbeat	Slows heartbeat
Arteries	Constricts arteries	Dilates arteries
Lungs	Dilates bronchial passages	Constricts bronchial passage
Digestive Tract	Slows peristalsis rate	Increases peristalsis rate
Eye	Dilates pupil	Constricts pupil
Urinary Bladder	Relaxes bladder	Constricts bladder

THE NERVOUS SYSTEM OF OTHER ORGANISMS

Protozoans – Protozoans have no nervous system; however, their protoplasm does receive and respond to certain stimuli.

Hydra – The hydra possesses a simple nervous system, which has no central control, known as a nerve net. A stimulus applied to a specific part of the body will generate an impulse which will travel to all body parts.

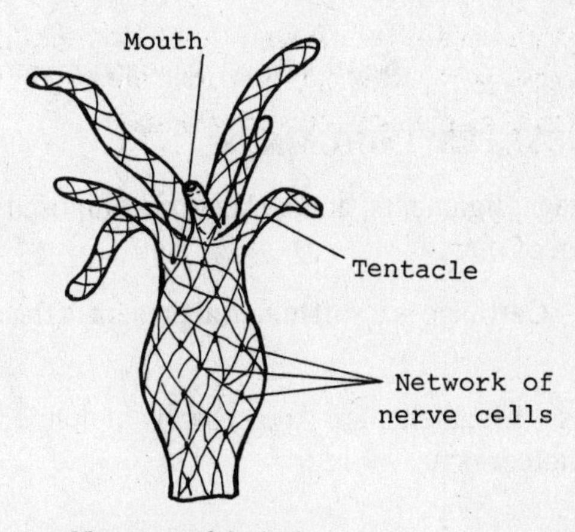

Nerve net of the hydra.

Earthworm – The earthworm possesses a central nervous system which includes a brain, a nerve cord (which is a chain of ganglions), sense organs, nerve fibers, muscles, and glands.

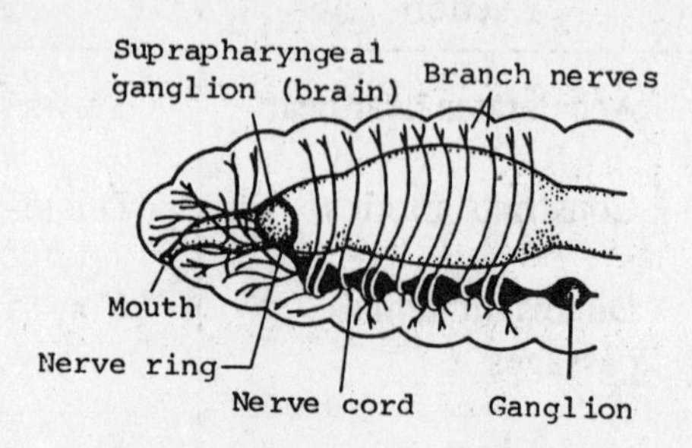

Nervous system of the earthworm.

Grasshopper – The grasshopper's nervous system consists of ganglia bundled together to form the peripheral nervous system. The ganglia of the grasshopper are better developed than in the earthworm.

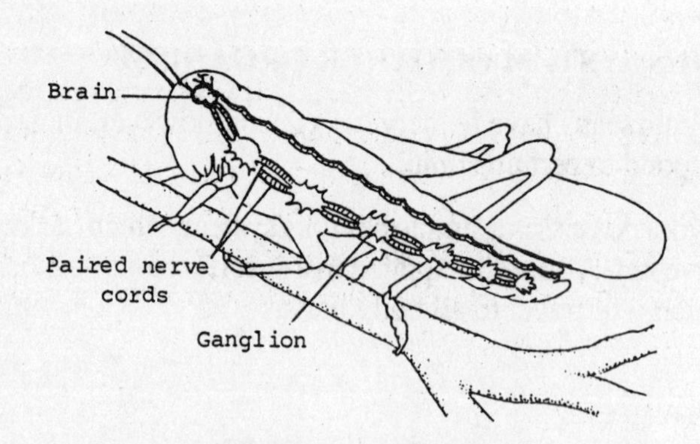

Nervous system of the grasshopper.

SKELETAL SYSTEM IN HUMANS

The cartilage, ligaments, and a skeleton composed of 206 bones make up the skeletal system of man.

A) Cartilage – Cartilage is a soft material present at the ends of bones, especially at joints.

B) Ligaments – Ligaments are strong bands of connective tissue which bind one bone to another bone.

C) Skeleton – The bones of the skeleton serve five important functions:

1) Allowing movements of parts of the body.
2) Supporting various organs of the body.
3) Supplying the body with red blood cells and some white blood cells.
4) Protecting internal organs.
5) Storing calcium and phosphate salts.

The skeleton is composed of the vertebral column, the skull, the limbs, the breastbone, and the ribs.

E. EVOLUTION

The Origin of Life

Soon after the earth's formation, when conditions were quite different from those existing today, a period of spontaneous chemical synthesis began in the warm ancient seas. During this era, amino acids, sugars, and nucleotide bases—the structural subunits of some of lifes' macromolecules—formed spontaneously from the hydrogen-rich molecules of ammonia, methane, and water. Such spontaneous synthesis was only possible because there was little oxygen in the atmosphere. The energy for synthesis came in the form of lightning, ultra-violet light, and higher energy radiations. These molecules polymerized due to their high concentrations, and eventually the first primitive signs of life arose.

Evidence for Evolution

The evidence for evolution can be explained by these eight facts:

A) Comparative Anatomy – Similarities of organs in related organisms show common ancestry.

B) Vestigial Structures – Structures of no apparent use to the organism, may be explained by descent from forms that used these structures.

C) Comparative Embryology – The embryo goes through developmental stages in common with other types of species.

D) Comparative Physiology – Many different organisms have similar enzymes. Mammals have similar hormones.

E) Taxonomy – All organisms can be classified into kingdom, phylum, class, order, family, genus, and species. This commonness in classification seems to indicate relationships between organisms.

F) Biogeography – Natural barriers, such as oceans, deserts, and mountains, which restrict the spread of species to other favorable environments. Isolation frequently produces many variations of species.

G) Genetics – Gene mutations, chromosome segment rearrangements, and chromosome segment doubling produce variations and new species.

H) Paleontology – Present individual species can be traced back to origins through skeletal fossils.

Natural Selection

The Darwin-Wallace theory of natural selection states that a significant part of evolution is dictated by natural forces, which select for survival those organisms that can respond best to certain conditions. Since more organisms are born than can be accommodated by the environment, a limited number is chosen to live and reproduce. Variation is characteristic of all animals and plants, and it is this variety which provides the means for this choice. Those individuals who are chosen for survival will be the ones with the most and best adaptive traits. These include the ability to compete successfully for food, water, shelter, and other essential elements; the ability to reproduce and perpetuate the species; and the ability to resist adverse natural forces, which are the agents of selection.

Factors Affecting Allele Frequencies: Genetic Drift, Bottleneck Effect, Founder Effect

Changes in allele frequency are caused by migration, mutation, and genetic drift. Genetic drift refers to the absence of natural selection. Random changes in gene frequencies occur, including the random loss of alleles. Random changes in the gene pool can be produced by catastrophic events where there are few survivors. Following such events, allele frequencies in the population may become quite different through chance alone. Such events are sometimes called population bottlenecks. A bottleneck effect can be created where a few people colonize a new territory. This is known as a founder effect.

Mechanisms of Speciation: Isolating Mechanisms, Allopatry, Sympatry, Adaptive Radiation

The following three mechanisms are fundamental in speciation:

A) Allopatric speciation is the formation of new species through the geographic isolation of groups from the parent population (as occurs through colonization or geological disruption).

B) Sympatric speciation occurs within a population and without geographical isolation. It is rare in animals, but not in plants.

C) Adaptive radiation is the formation of new species arising from a common ancestor resulting from their adaptation to different environments.

Evolutionary Patterns: Gradualism, Punctuated Equilibrium

Gradualism is the belief that evolution occurs very gradually, or in such small increments as to be nearly unobservable. In contrast, punctuated equilibrium describes evolutionary change occurring in relatively sudden spurts, following long periods of minor change or no change.

Behavior

LEARNED BEHAVIOR

Conditioning, habits, and imprinting are all specific types of behavior that are learned and acquired as the result of individual experiences.

A) Conditioning – Conditioned behavior is a response caused by a stimulus different from that which originally triggered the response. Experiments conducted by Pavlov on dogs demonstrate conditioning of behavior.

ringing of bell (stimulus 1) → barking (response 1)

food + ringing of bell (stimulus 2) → saliva flow (response 2)

ringing of bell (stimulus 1) → saliva flow (response 2)

Conditional behavior of Pavlov's dogs.

B) Habits – Habit behavior is learned behavior that becomes automatic and involuntary as a consequence of repetition. When an action is constantly repeated, the amount of thinking is reduced because impulses pass through the nerve pathways more quickly. The behavior soon becomes automatic.

C) Imprinting – Imprinting involves the establishment of a fixed pathway in the nervous system by the stimulus of the very first object that is seen, heard, or smelled by the particular organism. The research of Konrad Lorenz with newly-hatched geese demonstrated this type of learning.

INNATE BEHAVIOR

Taxis, reflexes, and instincts are all specific types of behavior that are inborn and involuntary.

A) Taxis – Taxis is the response to a stimulus by automatically moving either toward or away from the stimulus.

1) Phototaxis – Photosynthetic microorganisms move toward light of moderate intensity.

2) Chemotaxis – Organisms move in response to some chemical.

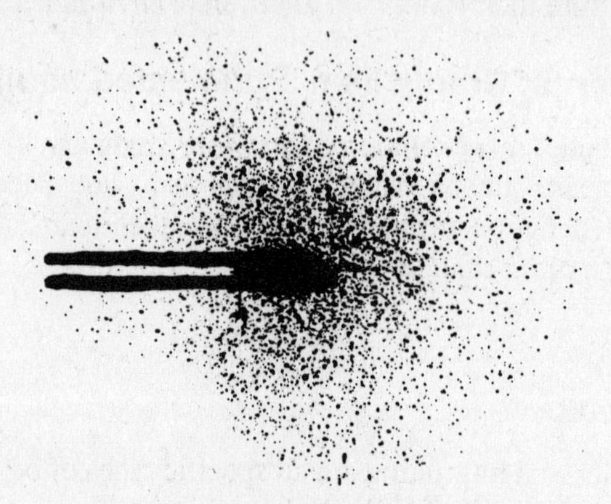

E. coli bacteria congregate near specific chemical.

B) Reflex – A reflex is an automatic response to a stimulus in which only a part of the body is involved; it is the simplest inborn response.

The knee jerk is a stretch reflex that is a response to a tap on the tendon below the knee cap which stretches the attached muscle. This tapping activates stretch receptors. Stretching a spindle fiber triggers nerve impulses.

C) Instinct – An instinct is a complex behavior pattern which is unlearned and automatic and is often beneficial in adapting the individual to its environment.

Nest-making of birds and web-spinning by spiders are examples of instinctive behavior.

1) Instinct of self-preservation – This is characterized by "fight or flight" behavior of animals.

2) Instinct of species-preservation – This is characterized by the instinctive behavior of the animal not to escape or fight, but to find a safer area for habitation.

3) Releasers – The releasers are signals which possess the ability to trigger instinctive acts.

VOLUNTARY BEHAVIOR

Voluntary behavior includes activities under direct control of the will, such as learning and memory.

A) Learning – Intelligence measures the ability to learn and properly establish new patterns of behavior. Humans demonstrate the highest degree of intelligence among all animals. This is due, in part, to the highly developed cerebrum which contains a great quantity of nerve pathways and neurons.

B) Memory – All learning is dependent upon one's memory. Memory is essential for all previous learning to be retained and used.

BIOLOGY

TEST I

BIOLOGY
ANSWER SHEET

1. (A) (B) (C) (D) (E)
2. (A) (B) (C) (D) (E)
3. (A) (B) (C) (D) (E)
4. (A) (B) (C) (D) (E)
5. (A) (B) (C) (D) (E)
6. (A) (B) (C) (D) (E)
7. (A) (B) (C) (D) (E)
8. (A) (B) (C) (D) (E)
9. (A) (B) (C) (D) (E)
10. (A) (B) (C) (D) (E)
11. (A) (B) (C) (D) (E)
12. (A) (B) (C) (D) (E)
13. (A) (B) (C) (D) (E)
14. (A) (B) (C) (D) (E)
15. (A) (B) (C) (D) (E)
16. (A) (B) (C) (D) (E)
17. (A) (B) (C) (D) (E)
18. (A) (B) (C) (D) (E)
19. (A) (B) (C) (D) (E)
20. (A) (B) (C) (D) (E)
21. (A) (B) (C) (D) (E)
22. (A) (B) (C) (D) (E)
23. (A) (B) (C) (D) (E)
24. (A) (B) (C) (D) (E)
25. (A) (B) (C) (D) (E)
26. (A) (B) (C) (D) (E)
27. (A) (B) (C) (D) (E)
28. (A) (B) (C) (D) (E)
29. (A) (B) (C) (D) (E)
30. (A) (B) (C) (D) (E)
31. (A) (B) (C) (D) (E)
32. (A) (B) (C) (D) (E)
33. (A) (B) (C) (D) (E)
34. (A) (B) (C) (D) (E)
35. (A) (B) (C) (D) (E)
36. (A) (B) (C) (D) (E)
37. (A) (B) (C) (D) (E)
38. (A) (B) (C) (D) (E)
39. (A) (B) (C) (D) (E)
40. (A) (B) (C) (D) (E)
41. (A) (B) (C) (D) (E)
42. (A) (B) (C) (D) (E)
43. (A) (B) (C) (D) (E)
44. (A) (B) (C) (D) (E)
45. (A) (B) (C) (D) (E)
46. (A) (B) (C) (D) (E)
47. (A) (B) (C) (D) (E)
48. (A) (B) (C) (D) (E)
49. (A) (B) (C) (D) (E)
50. (A) (B) (C) (D) (E)
51. (A) (B) (C) (D) (E)
52. (A) (B) (C) (D) (E)
53. (A) (B) (C) (D) (E)
54. (A) (B) (C) (D) (E)
55. (A) (B) (C) (D) (E)
56. (A) (B) (C) (D) (E)
57. (A) (B) (C) (D) (E)
58. (A) (B) (C) (D) (E)
59. (A) (B) (C) (D) (E)
60. (A) (B) (C) (D) (E)
61. (A) (B) (C) (D) (E)
62. (A) (B) (C) (D) (E)
63. (A) (B) (C) (D) (E)
64. (A) (B) (C) (D) (E)
65. (A) (B) (C) (D) (E)
66. (A) (B) (C) (D) (E)
67. (A) (B) (C) (D) (E)
68. (A) (B) (C) (D) (E)
69. (A) (B) (C) (D) (E)
70. (A) (B) (C) (D) (E)
71. (A) (B) (C) (D) (E)
72. (A) (B) (C) (D) (E)
73. (A) (B) (C) (D) (E)
74. (A) (B) (C) (D) (E)
75. (A) (B) (C) (D) (E)
76. (A) (B) (C) (D) (E)
77. (A) (B) (C) (D) (E)
78. (A) (B) (C) (D) (E)
79. (A) (B) (C) (D) (E)
80. (A) (B) (C) (D) (E)
81. (A) (B) (C) (D) (E)
82. (A) (B) (C) (D) (E)
83. (A) (B) (C) (D) (E)
84. (A) (B) (C) (D) (E)
85. (A) (B) (C) (D) (E)
86. (A) (B) (C) (D) (E)
87. (A) (B) (C) (D) (E)
88. (A) (B) (C) (D) (E)
89. (A) (B) (C) (D) (E)
90. (A) (B) (C) (D) (E)
91. (A) (B) (C) (D) (E)
92. (A) (B) (C) (D) (E)
93. (A) (B) (C) (D) (E)
94. (A) (B) (C) (D) (E)
95. (A) (B) (C) (D) (E)
96. (A) (B) (C) (D) (E)
97. (A) (B) (C) (D) (E)
98. (A) (B) (C) (D) (E)
99. (A) (B) (C) (D) (E)
100. (A) (B) (C) (D) (E)

BIOLOGY EXAM I

PART A

Directions: For each of the following questions or incomplete sentences, there are five choices. Choose the answer which is most correct. Darken the corresponding space on your answer sheet.

1. Light energy is captured in the plant by

(A) NADP.
(B) ATP.
(C) chlorophyll.
(D) glucose.
(E) carbon dioxide.

2. Pheromones are used by

(A) ants to mark trails to food.
(B) crickets to find each other for mating.
(C) bees to find nectar.
(D) birds in courtship.
(E) female frogs in recognizing males of the same species.

3. Bacteria have

(A) a cell wall.
(B) a nuclear membrane.
(C) chloroplasts.
(D) mitochondria.
(E) rod like chromosomes.

4. Intense physical activity can result in all of the following EXCEPT

(A) large amounts of glucose being converted to glycogen.
(B) the breakdown of glucose.
(C) an oxygen debt.
(D) ATP production.
(E) muscle fatigue.

5. Fungi have all of the following characteristics EXCEPT

(A) can undergo photosynthesis.
(B) have cell walls.
(C) live on dead organic matter.
(D) produce spores.
(E) secrete digestive enzymes.

6. The total area within which an animal travels during its routine activities and may share with other animals, but does not necessarily defend is its

(A) individual distance.
(B) home range.
(C) territory.
(D) migration area.
(E) nesting area.

7. Terrestrial, cold blooded animals that lay hard shelled eggs are

(A) fishes.
(B) reptiles.
(C) birds.
(D) amphibians.
(E) mammals.

8. The structures responsible for movement are

(A) cilia.
(B) ribosomes.
(C) chromosomes.
(D) Golgi apparatus.
(E) endoplasmic reticulum.

9.

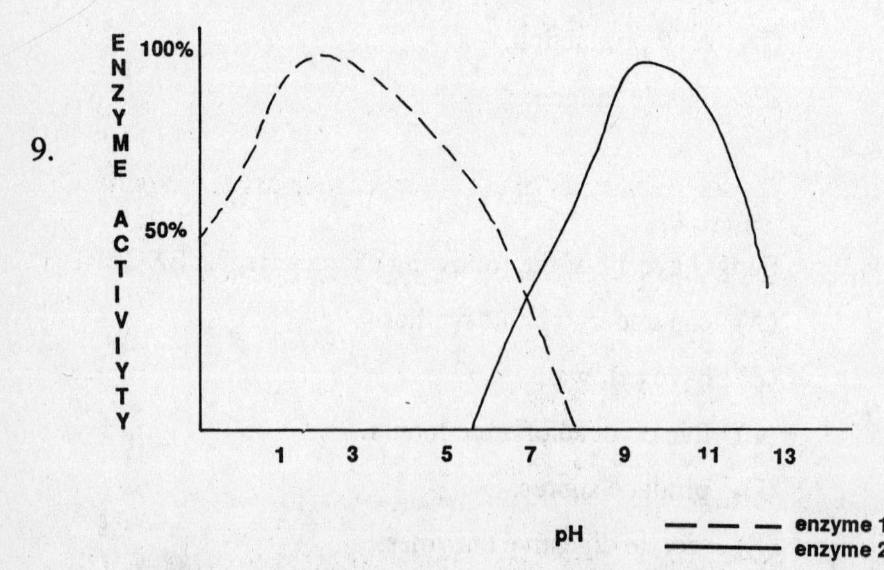

From the graph above showing the effect of pH on enzyme activity, which is correct?

(A) An enzyme has some activity at all pH values.
(B) One enzyme may have more than one optimum pH.
(C) The activity curve for an enzyme found in the stomach would be similar to enzyme 1.
(D) Both enzymes have an optimum activity at about pH 7.
(E) Enzyme 2 works best at high acidity.

10. In land plants water absorption occurs through the

(A) leaves.
(B) root hairs.
(C) buds.
(D) stem.
(E) trunk.

11. When an animal learns to make a strong association with another organism during a brief period early in development, this is called

(A) conditioning.
(B) imprinting.
(C) reinforcement.
(D) habituation.
(E) trial and error learning.

12. The most abundant plants living in temperate, terrestrial habitats are

(A) conifers.
(B) ferns.
(C) club mosses.
(D) flowering plants.
(E) horsetails.

13. Fertilization of the egg in humans normally occurs in the

(A) ovary.
(B) uterus.
(C) oviduct.
(D) urethra.
(E) vagina.

14. Monocots and dicots have which of the following in common?

(A) annual growth rings enclosing a pith.

(B) network of veins in leaves

(C) one cotyledon.

(D) well developed xylem.

(E) petals of flowers occur in threes or multiples of three.

15. The sources of genetic variation include all of the following EXCEPT

(A) mutation.

(B) natural selection.

(C) fertilization of egg by sperm.

(D) crossing over.

(E) meiosis.

16. Excess tissue fluid is returned to the blood by the

(A) respiratory system.

(B) lymphatic system.

(C) urinary system.

(D) circulatory system.

(E) digestive system.

17. The movement of sodium ions across the cell membrane during a nerve impulse occurs by

I diffusion

II osmosis

III active transport

(A) I only
(B) II only
(C) III only
(D) I and II
(E) I and III

18. A dorsal hollow nerve cord is present in
(A) earthworms.
(B) grasshoppers.
(C) starfish.
(D) frogs.
(E) planaria.

19. Which of the following is NOT true about the synapse?
(A) neurotransmitter is released
(B) it may be part of a reflex arc
(C) integration of information can occur here
(D) it includes a space between neurons
(E) it occurs primarily in the peripheral nervous system

20. The term primary productivity refers to which part of the food chain?
(A) herbivores
(B) carnivores
(C) omnivores
(D) plants
(E) decomposers

21. Nutrient absorption occurs primarily in the

(A) esophagus.
(B) stomach.
(C) small intestine.
(D) large intestine.
(E) rectum.

22.

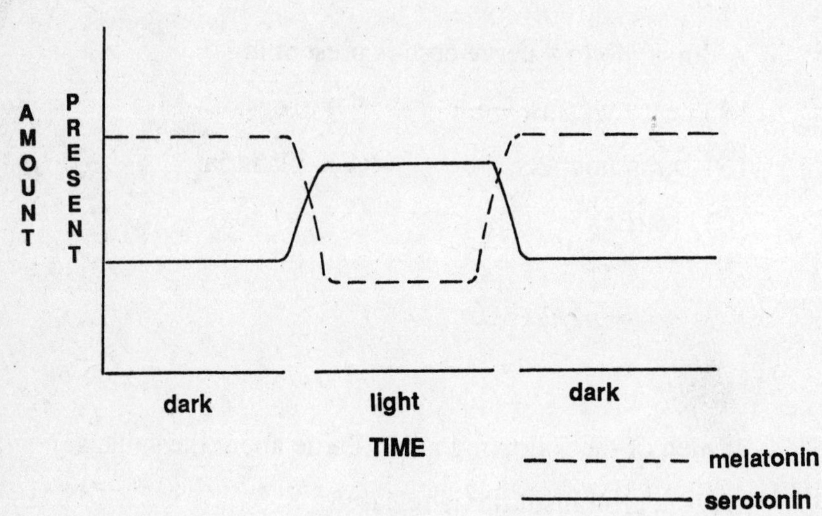

The graph above shows the production of serotonin (a precursor of melatonin) and melatonin (a hormone made by the pineal gland) during a 24 hour period. From this graph, which of the following is true?

(A) light increases melatonin production and increases serotonin production.

(B) light increases melatonin production and inhibits serotonin production.

(C) light inhibits melatonin production and inhibits serotonin production.

(D) light inhibits melatonin production and increases serotonin priduction.

(E) Serotonin inhibits melatonin production.

23. Which of the following groups is most closely related?

(A) ciliates - sponges

(B) flatworms - roundworms

(C) sea anemones - flagellates

(D) segmented worms - insects

(E) clams - squid

24. Which of the following is NOT normally filtered out of the blood in the kidney?

(A) protein

(B) glucose

(C) sodium chloride

(D) water

(E) urea

25. Which of the following are NOT part of innate behavior?

(A) kineses

(B) reflexes

(C) insight

(D) sign stimuli

(E) taxes

26. Sperm maturation and storage occurs in the

(A) testes.

(B) urethra.

(C) vas deferens.

(D) epididymus.

(E) ureter.

27. If a red-eyed female fruit fly who carries a recessive gene for white eyes on her X chromosome mates with a red-eyed male, what will be expected among the offspring?

(A) 1/2 will have white eyes

(B) 3/4 red eyes, 1/4 white eyes for both sexes

(C) all the males will have white eyes

(D) 1/2 the males will have white eyes

(E) white eyes will only show among the females

28. Nitrogenous waste removal from the chick embryo occurs by

(A) diffusion across the placenta.

(B) imaginal discs.

(C) the allantois.

(D) diffusion across the eggshell.

(E) the yolk sac.

29. A nerve has the ability to repair itself unless which is destroyed?

(A) dendrites

(B) cell body

(C) axon

(D) terminal branches

(E) synapse

30. Bone to muscle connections are made by

(A) tendons.

(B) ligaments.

(C) bone.

(D) cartilage.

(E) striated muscle.

31. Natural selection depends upon all the following EXCEPT

(A) more individuals are born in each generation than will survive and reproduce.

(B) There is genetic variation among individuals.

(C) Some individuals have a better chance of survival than others.

(D) Some individuals are reproductively more successful.

(E) Certain acquired traits can be passed on to the next generation.

32. Which of the following does NOT match?

(A) aldosterone - water retention

(B) thyroxin - metabolism

(C) FSH - maturation of ovum

(D) ADH - growth

(E) adrenalin - respiratory rate

33. In the cross tall smooth x tall smooth the offspring were

9/16 tall smooth

3/16 tall wrinkled

3/16 short smooth

1/16 short wrinkled

From this we can conclude

(A) The genes for tall and smooth are linked.

(B) The genes for tall and wrinkled are dominant.

(C) At least one of the parents was homozygous for smooth.

(D) At least one of the parents was homozygous for tall.

(E) At least one of the parents was heterozygous for tall.

34. The cell membrane pores are made of

(A) fats.

(B) proteins.

(C) sugar.

(D) DNA.

(E) starches.

35. Which of the following situations is most likely to result in an evolutionary change?

(A) athletic ability by extensive training

(B) maintenance of health by correct diet

(C) prompt medical attention when exposed to disease

(D) increased resistance to a disease because of a mutation

(E) vitamin supplements

36. The liver does all of the following, EXCEPT

(A) makes bile.

(B) stores glucose.

(C) makes urea.

(D) converts glucose into fat.

(E) makes white blood cells.

37. Which of the following does NOT match?

(A) ribosomes - RNA production

(B) mitochondria - ATP formation

(C) Golgi apparatus - packaging of secretory products

(D) lysosomes - store digestive enzymes

(E) centrioles - formation of spindle fibers

38. Radial symmetry is found in the

(A) Cnidarians.
(B) Platyhelminthes.
(C) Mollusca.
(D) Arthropoda.
(E) Annelida.

39. Which of the following is expected to have the largest mass of living material (biomass)?

(A) producers
(B) herbivores
(C) decomposers
(D) carnivores
(E) predators

40. Evidence for evolution includes all of the following EXCEPT

(A) fossil record
(B) similarities of proteins in different organisms
(C) homologous limb structures
(D) similarities in chromosome banding patterns
(E) differences in physical appearance of individuals within a species

41. In the spring, the ice on a lake melts. The cooler surface water settles to the bottom, displacing the warmer bottom water. All of the following will happen EXCEPT

(A) nutrients are brought to the surface
(B) phosphorus is made available to living organisms
(C) algae will increase in growth
(D) upwellings of bottom water will occur
(E) extremes in water temperature will form from top to bottom

42. The primary structural component of the cell wall is

(A) cellulose.
(B) auxin.
(C) phospholipids.
(D) proteins.
(E) ethylene.

43. An ecological community with permafrost is the

(A) tropical rain forest.
(B) taiga.
(C) tundra.
(D) temperate forest.
(E) grassland.

44. Which tissue has the ability to repair itself most rapidly?

(A) epithelium
(B) connective
(C) muscle
(D) nerve
(E) bone

45. Cockleburs were grafted together but remained separated by a light-tight barrier. They were exposed to a particular light cycle and the results are shown in the drawing below. From these results, which of the following is true?

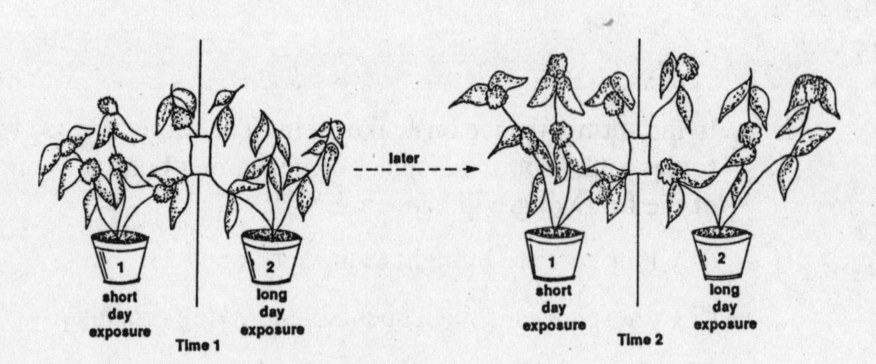

(A) Cockleburs are long day plants.

(B) Cockleburs are short night plants.

(C) A stimulus from plant 1 induces flowering in plant 2.

(D) Plant 2 inhibits flowering in plant 1.

(E) Day length has no effect on flowering.

46. The development of a secretory uterus in preparation for implantation is most directly under the control of

(A) FSH.

(B) LH.

(C) estrogen.

(D) progesterone.

(E) HCG.

47. All land plants have

(A) vascular tissue.

(B) a dominant sporophyte generation.

(C) seeds.

(D) a cuticle on parts exposed to air.

(E) freely swimming gametes.

48. Which of the following statements is NOT true about human nutrition?

(A) Our bodies make some of the amino acids we require for growth.

(B) Fats are not a necessary part of our diets.

(C) Some vitamins are water soluble and are not retained by the body.

(D) We must have all the amino acids in our body at the same time in order to make proteins.

(E) Fat soluble vitamins can be harmful when taken in excess.

49. DNA replication occurs immediately before

I mitosis
II meiosis 1
III meiosis 2

(A) I only
(B) II only
(C) III only
(D) I and II
(E) I, II and III

50. Assuming the environmental factors remain constant, which of the following successional stages represents the most stable community?

(A) a lake
(B) freshwater marsh
(C) meadow
(D) low shrubs
(E) forest

51. The gene frequencies in a population may change because of all of the following EXCEPT

(A) mutation
(B) migration
(C) selection
(D) nonrandom mating
(E) a stable environment

52. Which is NOT a correct match?

(A) ectoderm - nervous system
(B) mesoderm - muscle
(C) ectoderm - lens of the eye
(D) endoderm - skin
(E) mesoderm - bone

53. Cellular metabolism includes all of the following EXCEPT

(A) protein synthesis

(B) cellular respiration

(C) burning of fats

(D) conversion of sugars to fats

(E) transport of nutrients in the blood

54. Two plants were grown under identical conditions except that gibberellic acid was applied to one of them. Plant heights were measured and the following results were obtained.

	height in millimeters	
day	gibbrerellic acid	control
1	35	35
4	45	39
7	50	41
10	60	43
13	70	46

Based upon the results above, gibberellic acid

(A) causes an increase in cell division only.

(B) causes an increase in cell elongation only.

(C) causes an increase in both cell division and cell elongation.

(D) has no effect on growth.

(E) has an effect on growth, but we cannot tell from this experiment whether it is due to cell division and/or cell elongation.

55. When distinguishing between DNA and RNA, DNA uniquely contains

(A) guanine.

(B) ribose.

(C) phosphorus (phosphates).

(D) one strand.

(E) thymine.

56. Animal cells usually contain all of the following EXCEPT

(A) mitochondria.

(B) chloroplasts.

(C) ribosomes.

(D) Golgi bodies.

(E) chromosomes.

57. The oxygen that is involved in photosynthesis

(A) is an end product.

(B) is used to make ATP.

(C) is a raw material for glucose.

(D) captures the energy from sunlight.

(E) is needed for NADPH production.

58. The functions of meiosis might include any of the following EXCEPT

(A) reduce the chromosome number by half.

(B) produce many sperm.

(C) provide the egg with a large volume of cytoplasm.

(D) allow for crossing over.

(E) provide for identical offspring.

PART B

DIRECTIONS: Each of the following sets of questions refers to a list of lettered choices immediately preceding the question set. Choose the best answer from the list, then darken the corresponding space on your answer sheet. Choices may be used once, more than once, or not at all.

Questions 59 - 62

(A) carbohydrate

(B) fat

(C) protein

(D) deoxyribose nucleic acid

(E) ribose nucleic acid

59. Compound containing amino acids.

60. Compound that makes up genes in plants and animals.

61. Compound that is the end product of photosynthesis.

62. Compound that can be an enzyme.

Questions 63 - 64

(A) community
(D) ecosystem

(B) population
(E) niche

(C) habitat

63. A group of individuals belonging to the same species.

64. The way an organism makes its living.

Questions 65 - 68

(A) parasitism
(B) mutualism
(C) predation
(D) commensalism
(E) competition

65. A deer tick biting a deer is an example of

66. Two species of frogs colonize the same lake. One increases in number, while the other decreases in number.

67. Photosynthetic algae and water capturing fungi combine to form lichens. The algae provide nutrients and the fungi supply water.

68. Orchids grow on the branches of trees. The orchids are held above ground level where the light is better. The tree is neither helped nor harmed by the orchid.

Questions 69 - 70

(A) binary fission
(B) rhizome
(C) mitosis
(D) budding
(E) meiosis

69. Asexual reproduction in yeast.

70. Occurs in prokaryotes.

Questions 71 - 74

(A) gastrulation
(B) cleavage
(C) implantation
(D) blastula formation
(E) neurulation

71. The formation of the central nervous system.

72. Digestive tract formation.

73. Triggered by fertilization.

74. Initiated by trophoblast secretions.

Questions 75 - 78

(A) messenger RNA
(B) gene
(C) ribosome
(D) nucleus
(E) transfer RNA

75. Site of protein synthesis.

76. Amino acid transport.

77. Can mutate.

78. Directs the assembly of amino acids into proteins

PART C

DIRECTIONS: The following questions refer to the accompanying diagram. Certain parts of the diagram have been labelled with numbers. Each question has five choices. Choose the best answer and darken the corresponding space on your answer sheet.

Questions 79 - 83

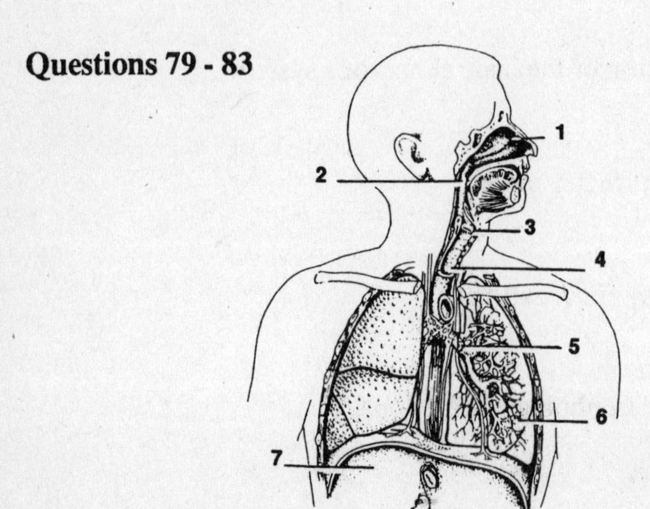

79. Exchanges of gases occur at

(A) 1
(B) 2
(C) 4
(D) 5
(E) 6

80. Voice production occurs in

(A) 1
(B) 2
(C) 3
(D) 4
(E) 5

81. The inspiration of air is caused by

(A) 1
(B) 2
(C) 4
(D) 6
(E) 7

82. Cilia and mucus trap debris at

(A) 2
(B) 3
(C) 4
(D) 6
(E) 7

83. Contains cartilage as C-shaped rings

(A) 1
(B) 2
(C) 4
(D) 6
(E) 7

PART D

DIRECTIONS: Each of the following questions refer to a laboratory or experimental situation. Read the description of each situation, then select the best answer to each question. Darken the corresponding space on the answer sheet.

Questions 84 - 86

Two species of flour beetles were grown together. In one case they were grown in pure flour. In the second case, they were grown under the same conditions, including the same amount of flour except that glass tubing was added.

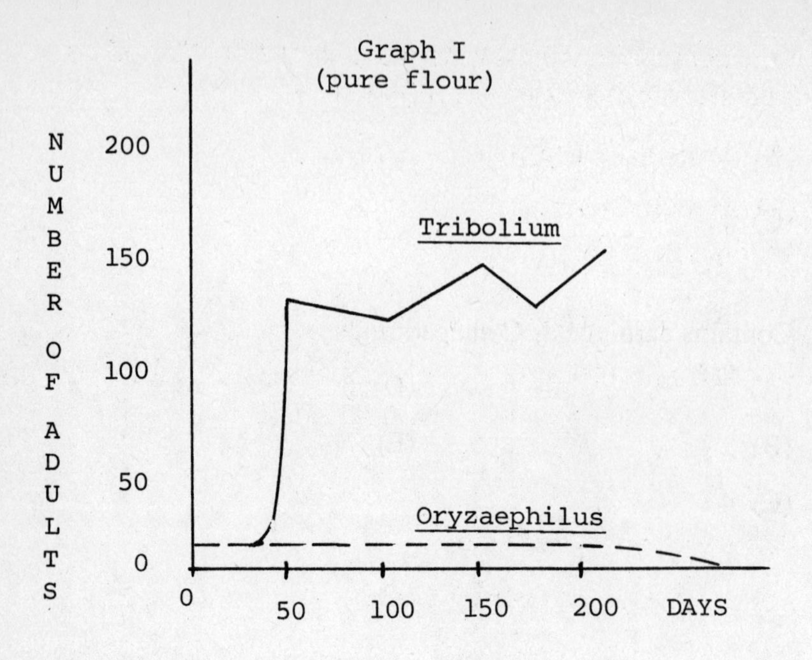
Graph I
(pure flour)
NUMBER OF ADULTS
200
150
100
50
0
Tribolium
Oryzaephilus
0
50
100
150
200
DAYS

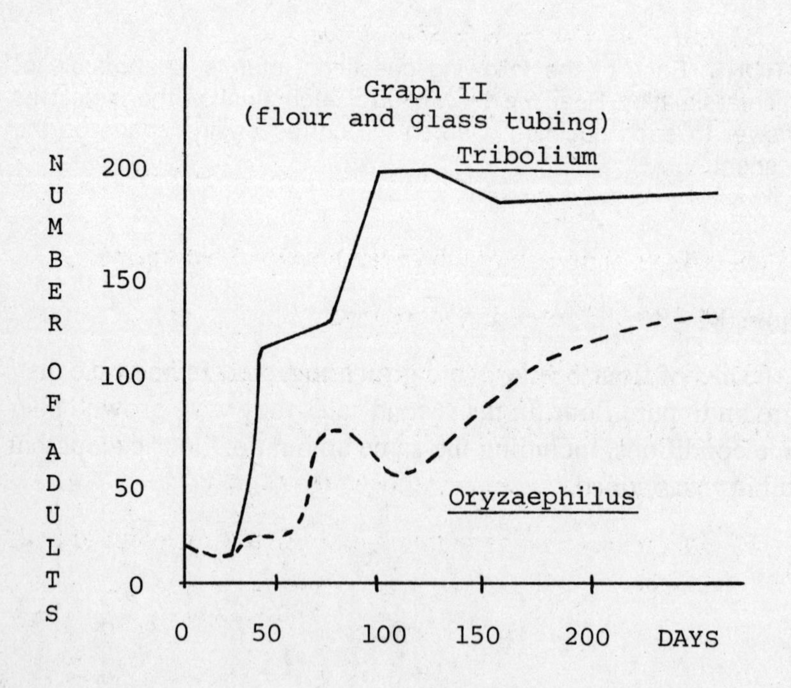
Graph II
(flour and glass tubing)
NUMBER OF ADULTS
200
150
100
50
0
Tribolium
Oryzaephilus
0
50
100
150
200
DAYS

84. When grown in pure flour, the number of *Tribolium* adults increases at the greatest rate between

(A) day 0 and 25 days

(B) 25 and 50 days

(C) 50 and 100 days

(D) 100 and 150 days

(E) 150 and 200 days

85. The presence of glass tubing

(A) inhibits growth in both species

(B) improves the growth of *Oryzaephilus*, but does not affect *Tribolium*

(C) improves the growth of *Tribolium*, but does not affect *Oryzaephilus*

(D) improves the growth of both species

(E) causes the extinction of *Oryzaephilus*

86. The effects shown in graph II can be explained as due to

(A) *Oryzaephilus* eating more flour

(B) competition between *Oryzaephilus* and *Tribolium*

(C) *Oryzaephilus* eating the *Tribolium* larvae

(D) changes in the temperature of the flour

(E) the presence of more than one habitat for growth of both species

Questions 87 - 90

	total cross-sectional area	velocity	blood pressure
aorta	4 cm^2	40 cm/sec	100 mmHg
arteries	20	10	80
capillaries	4000	0.05	20
veins	50	4	10
vena cava	12	10	8

87. Which part of the circulatory system displays the slowest blood flow?

(A) aorta
(B) arteries
(C) capillaries
(D) veins
(E) vena cava

88. Which part of the circulatory system allows for exchanges of nutrients, gases, and waste products between the blood and tissues?

(A) pulmonary arteries
(B) systemic arteries
(C) capillaries
(D) systemic veins
(E) pulmonary veins

89. Which of the following statements is true?

(A) Blood pressure increases as cross-sectional area decreases.

(B) Blood pressure decreases as cross-sectional area decreases.

(C) As cross-sectional area increases velocity increases.

(D) As cross-sectional area increases velocity decreases.

(E) As velocity increases blood pressure increases.

90. Blood that enters the aorta flows from the

(A) pulmonary arteries
(B) right atrium
(C) left atrium
(D) right ventricle
(E) left ventricle

Questions 91 - 92

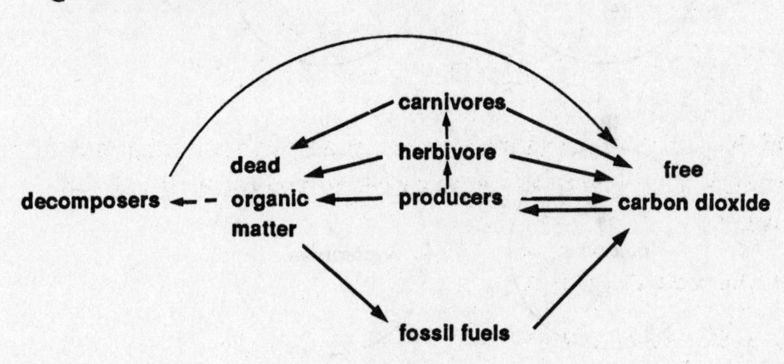

91. The building of carbon compounds occurs by

(A) photosynthesis
(B) respiration
(C) combustion
(D) decomposition
(E) death

92. Which of the following organisms would be a decomposer?

(A) oak tree
(B) mushroom
(C) caterpillar
(D) horse
(E) lion

Questions 93 - 94

A particular strain of bacteria does not have the ability to make the amino acid histidine because of a defective gene. If histidine is supplied in the medium, the bacteria will grow normally, as shown on plate I. When grown on minimal medium, few, if any, colonies grow (plate II). This strain was grown on two petri plates containing minimal medium. Chemical A was added to plate III and chemical B was added to plate IV. All the plates were grown at 37° C with colony counts made at the end of 48 hours.

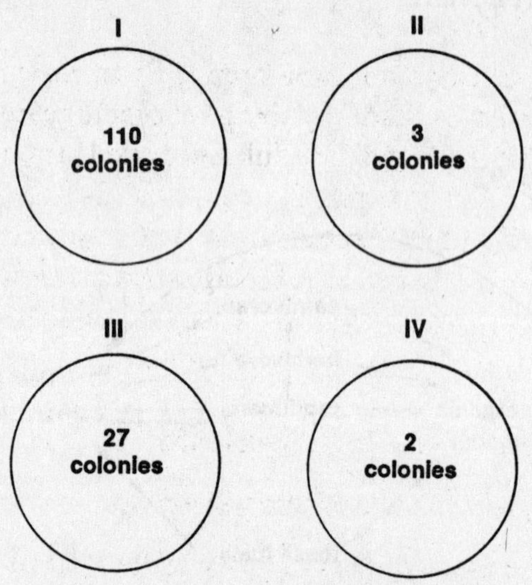

93. Which of the following statements about the experiment above is correct?

(A) Petri dishes I and III are the experimental sets.
(B) Dish I is the control for dish III.
(C) Exposure to the chemicals is the variable.

(D) The colony count is the variable.

(E) Temperature is the variable.

94. The results of this experiment indicate that

(A) chemical A has no effect upon survival of cells

(B) some bacteria have become sensitive to complete medium

(C) bacteria are always killed by minimal medium

(D) chemical A has caused mutations of the defective histidine gene

(E) chemical B has caused mutations of the defective histidine gene

Questions 95 - 96

Two hormones control metamorphosis in insects. Ecdysone, or molting hormone, is made by the prothoracic gland and causes the formation of the pupa and the adult. Juvenile hormone is made by the corpus allatum and prevents the ecdysone from working. When an active corpus allatum and prevents the ecdysone from working. When an active corpus allatum from a young larva is transplanted into a larva that is full grown, the second larva grows into a giant larva.

95. The results from this indicate that

(A) less ecdysone is being made

(B) the age of the larva has an effect on the amount of ecdysone made

(C) the age of the corpus allatum has an effect on the amount of juvenile hormone made

(D) the amount of ecdysone does not change, but more juvenile hormone is produced as the larva gets older

(E) the corpus allatum produces ecdysone

96. If the prothoracic gland were removed from a young larva, we would expect the larva to

(A) stop growing.

(B) grow to full size, but not pupate.

(C) grow to full size and pupate.

(D) pupate immediately.

(E) become a giant larva.

Questions 97 - 100

The diameter of sea urchin eggs was measured before they were placed in different concentrations of salt water. After 5 minutes their diameters were again measured.

Concentration of salt water	Percent change in diameter of eggs
0.0%	+20%
0.5	+10
2.5	+ 5
3.5	0
5.5	-20

97. Which of the following statements is true?

(A) The cells in the highest salt concentration enlarged the most.

(B) The cells in salt free water enlarged the most.

(C) All the cells changed uniformly.

(D) No water crossed the cell membrane in the cells in 3.5% salt concentration.

(E) 2.5% salt solution is isotonic to the cells.

98. Which salt water concentration is closest to the natural environment of this organism?

(A) 0%
(B) 0.5%
(C) 2.5%
(D) 3.5%
(E) 5.5%

99. The change in size is because of

(A) the movement of salt across the cell membranes.
(B) the movement of salt and water across the cell membranes.
(C) the movement of water across the cell membranes
(D) the growth of the cells.
(E) the loss of cell proteins.

100. If plant cells were placed in the same salt concentrations what would happen?

(A) The cells in highest salt concentration would enlarge the most.
(B) The cells in salt free water would burst.
(C) All the cells would change uniformly.
(D) No water would cross the cell membrane in the cells in 0.5% salt concentration.
(E) The cells in salt free water would be most firm.

BIOLOGY
TEST I

ANSWER KEY

1. C
2. A
3. A
4. A
5. A
6. B
7. B
8. A
9. C
10. B
11. B
12. D
13. C
14. D
15. B
16. B
17. E
18. D
19. E
20. D
21. C
22. D
23. E
24. A
25. C
26. D
27. D
28. C
29. B
30. A
31. E
32. D
33. E
34. B
35. D
36. E
37. A
38. A
39. A
40. E
41. E
42. A
43. C
44. A
45. C
46. D
47. D
48. B
49. D
50. E
51. E
52. D
53. E
54. E
55. E
56. B
57. A
58. E
59. C
60. D
61. A
62. C
63. B
64. E
65. A
66. E
67. B
68. D
69. D
70. A
71. E
72. A
73. B
74. C
75. C
76. E
77. B
78. A
79. E
80. C
81. E
82. C
83. C
84. B
85. D
86. E
87. C
88. C
89. D
90. E
91. A
92. B
93. C
94. D
95. C
96. B
97. B
98. D
99. C
100. E

BIOLOGY
TEST I

DETAILED EXPLANATIONS OF ANSWERS

1. (C)

Chlorophyll is the pigment that absorbs light energy, raising electrons to a higher energy level. These excited electrons transfer their energy to chemical bond energy. NADP has the ability to pick up chemical bond energy and ATP stores chemical bond energy. Carbon dioxide is the source of carbon for making glucose, the end product of photosynthesis.

2. (A)

For this question you must be familiar with behavioral interactions involving communication. Mate recognition by crickets and frogs is auditory. Scouting bees find nectar and communicate its location to the hive by dances. Birds have visual and auditory displays in courtship. Only the ants use chemical secretions or pheromones to mark their trails for other ants to find food.

3. (A)

Because bacteria are prokaryotes, they do not have a distinct nucleus separated by a nuclear membrane. In addition, their chromosome is circular. They have neither chloroplasts nor mitochondria. However, they have a cell wall, which is a complex of proteins and carbohydrates.

4. (A)

This question tests your understanding of intermediary metabolism. Glycogen is converted to glucose to supply the large amounts of energy required by intense physical activity. Because glycolysis can take place more rapidly than oxidative phosphorylation, its products (pyruvic acid and NADH) are produced faster than they can be used by the mitochondria. Instead of accumulating they react to produce lactic acid. This allows NAD, which is in limited supply, to recycle and glycolysis can continue. As lactic acid builds up, the acidity of the muscle increases until the muscle can no longer function. At that point muscle fatigue occurs. To process the lactic acid buildup, the oxygen debt must be paid.

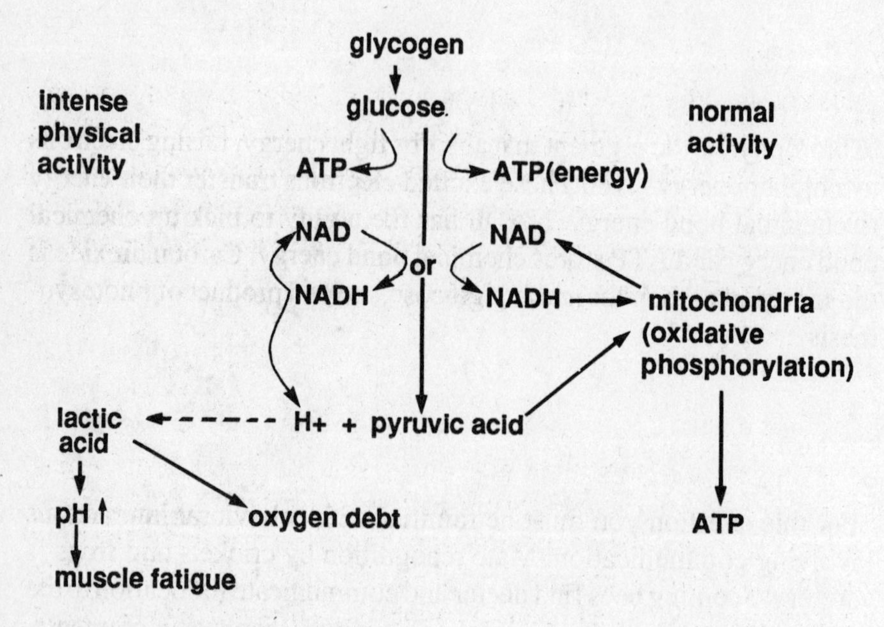

5. (A)

This question asks about the characteristics of fungi. Fungi are eukaryotes. They are multicellular and multinucleate. Structually they have cell walls that are composed mainly of chitin, a derivative polysaccharide containing nitrogen, and reproduce by spore production. They are not capable of undergoing photosynthesis, but instead feed on dead organic matter by secreting digestive enzymes into their environment, and breaking down food material extracellularly. The products of digestion are then absorbed through the cell wall and cell membranes by structures called haustoria.

6. (B)

Nesting area, individual distance, and territory are all aggressively defended against outside intruders. Both nesting areas and individual distances are actually small territories, the first being the size of the nest and the second being the immediate area around an animal. Migration involves the traveling to avoid adverse environmental conditions and/or to spawn. This includes the twice yearly journey of birds to avoid the harsh winters as well as the once in a lifetime journey of salmon to spawn. The routine travel in a home range is for food gathering and finding a mate, and does not necessarily include defense of the area.

7. (B)

Fishes are cold blooded, but are aquatic. Mammals and birds are warm blooded. While amphibians are cold blooded and may inhabit land, they lay eggs that have no amnion or shell. Thus, amphibians are tied to aquatic environments for the development of their offspring. Reptiles are the first vertebrate class that is truly terrestrial, because of their ability to lay hard shelled eggs.

8. (A)

Protein synthesis occurs on ribosomes, genes are found on chromosomes, and secretory products are packaged by the Golgi apparatus. Among other functions, the endoplasmic reticulum provides for intracellular transport for enzyme activity and for vesicle formation. Cilia are responsible for movement, because of the arrangement and interaction of microtubules within them.

9. (C)

The enzymes shown have specific ranges of activity and do not function at all pH values. Each enzyme has only one optimum pH as shown by the peak for each curve. The optimum for enzyme 1 is pH 2 and the optimum for enzyme 2 is pH 10. Enzyme 2 works best at low acidity, which is represented by a high pH value. Enzyme 1 works best at high acidity as indicated by a low pH. The enzyme activity curve for pepsin which is found in the acidic stomach has an optimum pH of about 2, thus, being most like enzyme 1.

10. (B)

Water loss occurs through the stomata of the leaves of land plants by transpiration. The outer leaf surfaces, buds, stems, and trunks of land plants have a waxy layer (cuticle) to prevent water loss. Root hairs do not have a cuticle and can absorb water by osmosis.

11. (B)

Conditioning is a type of learning that involves an association of a novel stimulus with a stimulus that is readily recognized. When the conditioning includes reward and punishment as part of the learning, it is called trial and error learning or reinforcement. Habituation is a gradual decline in response to sensation. Imprinting, on the other hand, involves the learning of a strong association that occurs during a brief period in development and is retained indefinitely.

12. (D)

Ferns, club mosses, and horsetails are land plants that are dependent upon a moist environment, because the sperm must swim to the egg for fertilization. Therefore, their abundances are limited. Conifers are the most abundant larger plants in the colder latitudes, while flowering plants dominate temperate (middle latitude) regions.

13. (C)

Fertilization is the union of the egg and sperm to form a zygote. The egg is formed and released from the ovary during ovulation. It is picked up and begins to travel down the oviduct. In humans, sperm are deposited in the vagina and swim through the uterus. The sperm continue to swim up the oviduct to the vicinity of the egg, and then fuse with the egg. The urethra is the duct leading from the bladder to the outside.

14. (D)

Monocots have one cotyledon and flower petals that occur in threes or multiples of three. Dicots have vascular bundles arranged as annual growth rings and leaves with net venation. As an adaptation to land, both types of flowering plants have a well developed xylem for transporting water.

15. (B)

This question tests your understanding of evolution and, in particular, knowing what will increase genetic variation. Mutations may involve changes in particular genes that result in new forms of the gene being produced. Meiosis is a process that involves producing gametes with one of each kind of chromosome present. The chromosomes for each egg or sperm are selected at random and, thus, result in each gamete having its own combination of chromosomes. During meiosis crossing over may occur between homologous chromosomes. This allows new combinations of genes to be located near each other and inherited as a unit on the new chromosome. The process of fertilization also increases genetic variability, since each egg and sperm contains its own combination of chromosomes. Natural selection is not a source of genetic variation, but rather acts on the genetic variation that already exists. Natural selection could, in fact, lead to decreased genetic variability.

16. (B)

The respiratory system provides for gas exchanges, intake of oxygen and release of carbon dioxide. The urinary system filters out body wastes. The digestive system breaks down food molecules into their components for absorption by the cells of the body. The circulatory system transports food, oxygen, and waste products throughout the body. As the blood circulates through the body, some of the plasma seeps out and bathes the cells of the body, providing for exchanges of materials with these cells. Tissue fluid that is not returned to the capillaries is picked up by the lymphatic vessels that run parallel to the circulatory system. These vessels drain into veins in the circulatory system.

17. (E)

Sodium is found in higher concentrations on the outside of the cell membrane. When the nerve is stimulated, the membrane becomes more permeable to sodium, and the sodium moves along the concentration gradient into the cell. Thus, the beginning of a nerve impulse is marked by sodium diffusing into the cell.

Na^+ (outside +)	(outside -)
K^+ (inside -)	Na^+ and K^+ (inside +)

At the end of a nerve impulse, the positive charge has been restored to the membrane. However, sodium that diffused into the cell must be removed and the potassium which diffused out must be returned to the inside of the cell.

K^+ (outside +)

Na^+ (Inside -)

The sodium is pumped out and the potassium is pumped in, using energy in the form of ATP. Thus, the end of the action potential is marked by a restoration of the sodium to its proper place by active transport. Therefore, sodium moves across the cell membrane during a nerve impulse by both diffusion and active transport.

18. (D)

This question highlights some features of the development of the vertebrate nervous system. The spinal cord forms on the dorsal surface and folds into the embryo to become a hollow nerve cord. Frogs are vertebrates and form this type of nerve cord. Earthworms, grasshoppers, and planaria have a pair of ventral solid nerve cords. The starfish has a ring of nervous tissue around its mouth and five nerves which extend into the arms along the ventral side.

19. (E)

The synapse is a structure that neurons use to communicate with each other. Information from the first neuron is transmitted to the second neuron via a chemical, the neurotransmitter. Since information may be received from sensory neurons and transmitted to motor neurons triggering an involuntary motor response, the synapse may be part of a reflex arc. Many neurons are involved in a synapse, and therefore, there is the opportunity to integrate information from a variety of neural sources. Most of the synapses are found in the central nervous system where most of the neurons are found and most integration of information occurs.

20. (D)

This question relates to energy flow in a food chain. Primary productivity is the amount of energy from the sun that is converted to chemical bond energy during photosynthesis. Plants are the only members of the ecosystem that can accomplish this conversion. Energy can then move to other organisms (herbivores, carnivores, omnivores, decomposers) through the food chain.

21. (C)

For this question, you must be familiar with the functions of the various parts of the digestive tract. Absorption of most digested materials and water occurs in the small intestine. Final water reabsorption occurs in the large intestine. The esophagus functions in the transport of food to the stomach, the stomach in the digestion of proteins, and the rectum in the storage of fecal matter.

22. (D)

The graph shows the amounts of melatonin and serotonin present during a 24 hour period. During the time of light exposure, melatonin production decreases and serotonin production increases, making (D) the correct choice. Furthermore, serotonin increases melatonin production in the dark, since serotonin is a precursor of melatonin.

23. (E)

This is a question about systematics. Four of the pairs of organisms represent different phyla:

ciliates (Ciliata) - sponges (Porifera)
flatworms (Platyhelminthes) - roundworms (Aschelminthes)
hydra (Cnidaria) - flagellates (Mastigophora)
segmented worms (Annelida) - insects (Arthropoda)

Only the clams and squid are members of the same phylum (Mollusca) and, therefore, are most closely related.

24. (A)

For this question an understanding of the functioning of the kidney is important. Some of the components of the blood that enter the glomerulus are filtered out into the nephron. These components are smaller in size and include glucose, sodium chloride, water, and urea. Protein is too large to pass through the glomerulus into the nephron tubule under normal circumstances.

25. (C)

Innate behavior is a genetically determined action performed by an animal at the right age given the right circumstances. Kineses are rapid, random body movements and taxes are directed body movements in response to environmental stimuli. Reflexes are rapid involuntary responses by a part of the body, and sign stimuli are releasers for stereotyped behavior. Kineses, taxes, reflexes, and sign stimuli are all examples of innate behavior. On the other hand, insight is a complex type of learning involving reasoning.

26. (D)

Sperm formation occurs in the testes. The sperm then travel to the epididymus for storage and maturation. The vas deferens provides fluid for the formation of semen which mixes with the sperm before ejaculation. The sperm are released through the urethra which travels through the penis. The ureter is part of the urinary system connecting the bladder and kidney.

27. (D)

For this question you must understand how sex linked traits are inherited. The female fruit fly is heterozygous for eye color and produces eggs that carry either an X containing a gene for red eyes or an X containing a gene for white eyes. The male produces sperm containing either an X containing a gene for red eyes or a Y chromosome that does not have an eye color gene. The Punnett square below indicates that all the female offspring will have red eyes, because they receive the dominant red eye gene from the father. The male offspring will have either red eyes (having received the red eye gene from the mother and a Y from the father) or will have white eyes (having received the white eye gene from the mother and a Y from the father).

<u>Key</u> R = red eye gene
r = white eye gene
Y = Y chromosome

<u>Cross</u>: red eye female x red eye male
Rr RY

<u>Eggs</u>: 1/2 R 1/2 r <u>Sperm</u>: 1/2 R 1/2 Y

<u>Punnett square</u>:

sperm

		1/2 R	1/2 Y
eggs	1/2 R	1/4 RR	1/4 RY
	1/2 r	1/4 Rr	1/4 rY

Results:

genotypes	phenotypes
RR	
Rr	
RY	red eyed females red eyed males
rY	white eyed males

28. (C)

The allantois provides nitrogenous waste removal. Diffusion of carbon dioxide occurs across the eggshell. However, the yolk sac is a nutrient source, rather than a site for waste removal. The placenta is an embryonic structure found in mammals and imaginal discs are found in insect larvae.

29. (B)

For this question you need to know the role of the nucleus in protein synthesis. Repair of damage to a cell depends upon production of new protein. Therefore, damage to the nerve is possible as long as the nucleus is intact and functioning.

30. (A)

This question requires a sense of the anatomical relationships of the portions of the body devoted to motion. Tendons connect bone to muscle, while ligaments provide bone to bone connections at the joint.

31. (E)

This question deals with the assumptions upon which evolution is based. Certain individuals have a better chance of survival, because of their genetic makeup. Since more individuals are born than can survive and reproduce, those individuals with the genetic advantage will have greater reproductive success. Acquired traits will not be passed on to the next generation, because there is no genetic basis for them.

32. (D)

This question relates particular hormones to their functions. The incorrect match is ADH and growth. ADH (antidiuretic hormone or vasopressin) increases water reabsorption by the kidney. Growth is primarily under the control of growth hormone, although other hormones (such as thyroxin) may indirectly affect growth.

33. (E)

Because short and wrinkled are traits that show up in the offspring and were not present in the parents, they must be caused by recessive genes. Therefore, the traits tall and smooth must be determined by dominant genes. In addition, both parents must have been heterozygous for both tall and smooth. Finally, the genes for height and texture assort independently as indicated by the 9:3:3:1 ratio and, they must, therefore, be on separate chromosomes.

<u>Key</u>: T = tall
t = short
S = smooth
s = wrinkled

<u>Cross</u>: TtSs x TtSs

<u>Gametes</u>: 1/4 TS 1/4Ts 1/4 tS 1/4 ts (same for both parents)

<u>Punnett square</u>:

	TS	Ts	tS	ts
TS	TTSS	TTSs	TtSS	TtSs
Ts	TTSs	TTss	TtSs	Ttss
tS	TtSS	TtSs	ttSS	ttSs
ts	TtSs	Ttss	ttSs	ttss

<u>Results</u>:

9/16 T_S_ - tall smooth

3/16 T_ss - tall wrinkled

3/16 ttS_ - short smooth

1/16 ttss - short wrinkled

34. (B)

Since the membrane is primarily a lipid bilayer, it is important that pores be present for the transport of large, charged particles. These pores are protein lined structures that enhance the ability of materials, such as glucose and salts, to cross the membrane.

35. (D)

This question focuses on one of the basic requirements for evolutionary change to occur. If change is to occur there must be genetic variation for natural selection to act upon. Athletic training, correct diet, medical care, and vitamin supplements may all improve the physical well being of the individual, but the improvements would not be passed on to the next generation. Mutation provides the raw material for evolution.

36. (E)

The liver makes bile, which is then stored in the gall bladder. It is also responsible for a variety of metabolic conversions. These include conversion of glucose and glycogen and storage of glycogen; conversion of nitrogenous wastes to urea; and conversion of excess glucose to fat. White blood cells are manufactured in bone marrow or lymphoid tissue, not the liver.

37. (A)

For this question you must know the functions of the organelles. The ribosomes are the site of protein synthesis. They contain RNA, but RNA production occurs in the nucleus under the direction of DNA.

38. (A)

For this question you must be familiar with the various invertebrates, and in particular with body design. The Cnidarians represent one of the earlier phyla as reflected by their more primitive body design (i.e. radial symmetry). Platyhelminthes, Mollusca, Arthropoda, and Annelida generally display bilateral symmetry.

39. (A)

Biomass is limited by the energy available to manufacture organic matter in living organisms. Energy enters the food chain via the producers who convert light energy to chemical bond energy. Since each transfer of organic material involves a loss of energy, each successive level of the food chain has less energy than the preceeding level. Therefore, the first level has the most energy and, thus, the largest amount of organic material or biomass.

40. (E)

This question requires some knowledge of the experimental data that support the concept of evolution. The more traditional evidence includes the fossil record and similarities in limb structures in different organisms. More recent evidence comes from molecular studies that include similarities in proteins and chromosome banding patterns. Differences in physical appearance within a species merely reflect individual variation with no sense of whether these traits are acquired or inherited, or whether changes have occurred over time.

41. (E)

Familiarity with aquatic ecosystems is important in this question. Dead plants and animals settle to the bottom of lakes and undergo decomposition. Nutrients, such as nitrogen, phosphorus, etc. are released during this process of decay. In order to sustain life in the lake these nutrients must be recycled. Ice floats on water and, as it melts, this cold water becomes denser and settles to the bottom of the lake. The warmer bottom water moves to the top as an upwelling, carrying the nutrients with it. These nutrients support the growth of living organisms, especially plants. This recycling of water also prevents temperature stratification of the lake during the period of upwelling.

42. (A)

The cell wall is made of cellulose. Auxin and ethylene (plant hormones), as well as proteins and phospholipids are found in plants, but are not structural components of the cell wall.

43. (C)

Knowledge of the different kinds of ecosystems on Earth is necessary for answering this question. Permafrost is a layer of frozen ground that never thaws. It has ecological implications in terms of the types of plants that will grow on this type of land. Permafrost is found in the areas closest to the poles - - in the tundra.

44. (A)

Nerve and muscle tissues have the ability to repair damaged cells, but cell division is absent in these tissues after birth. Connective tissue (and bone is one type) can repair itself, but this is a slow process. Epithelial tissue lines many surfaces that are subject to abrasion. Therefore, it regularly undergoes cell division and has a rapid rate of mitosis.

45. (C)

From the results recorded at time 1, we see that day length has an effect on flowering. Exposure to a short day light cycle induces flowering, while long day exposure does not induce flowering. From the results recorded at time 2, we see that flowering is induced later in the plant exposed to the long day light cycle. Since flowering was not initially induced by long day exposure, we must conclude that plant 1 stimulated plant 2 to flower. Long day and short night are the same thing.

46. (D)

FSH from the pituitary stimulates maturation of the ovum and estrogen production by the ovary. Estrogen stimulates thickening of the uterus. At mid-cycle an LH surge from the pituitary triggers ovulation and changes the follicle into a corpus luteum. The corpus luteum then produces progesterone in addition to estrogen. This progesterone causes the uterus to develop glands in preparation for its secretory activity when an embryo implants. HCG is a hormone made by the placenta that stimulates the corpus luteum to continue to make estrogen and progesterone during pregnancy.

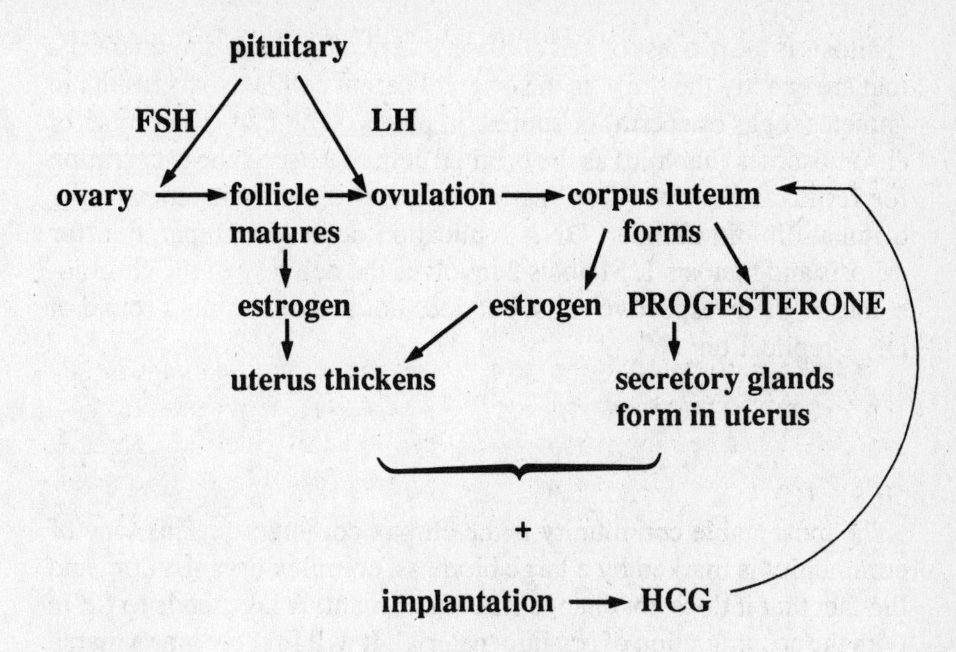

47. (D)

Bryophytes, the first land plants, lack vascular tissue, have a gametophyte generation that is dominant, and have swimming sperm. These land plants are, therefore, tied to moist environments for survival. While the nonseed tracheophytes have vascular tissue and a dominant sporophyte generation, they also have free swimming sperm. Seed bearing tracheophytes do not have free swimming sperm. The one characteristic that is common to all land plants is the presence of a cuticle on the parts exposed to the air. This protects them against desiccation.

48. (B)

While our bodies have the ability to make some of the raw materials needed for metabolic functions, some of these components must be ingested. These necessary components include certain amino acids, as well as certain fats. We must have all the amino acids in our bodies when protein synthesis is occurring, since proteins are assembled in a sequential manner and synthesis will stop if an amino acid is missing. Fat soluble vitamins are retained in the body and not excreted the way that water soluble vitamins are. Therefore, fat soluble vitamins can be taken in excess.

49. (D)

Mitosis is the process of cell division that results in two daughter cells that are exactly the same as the original parent cell. Meiosis results in gametes (eggs or sperm) or spores (in plants) with half the number of chromosomes (haploid) as the original cell. Meiosis is the preparation for fertilization, since fertilization returned the chromsomes to the original diploid number. DNA replication occurs in preparation for mitosis and meiosis 1. Meiosis 2 involves the delivery of the chromosomes to their respective cells, but does not involve another round of DNA replication.

50. (E)

The most stable community is the climax community. This type of community is marked by a large biomass, complex organization, and the fact that it does not change its environment. A lake tends to fill in with the accumulation of organic material. It will first become a marsh and then a meadow. A meadow provides open areas for the growth of shrubs, and trees then replace the shrubs.

51. (E)

This question relates to the Hardy-Weinberg equation, which deals with the gene frequencies of alleles in a stable population. Mutation, migration, selection, and nonrandom mating will all alter the frequency of certain genes by favoring some alleles over others. An unstable environment would also alter gene frequencies by favoring alleles that are now more beneficial in the new setting.

52. (D)

This question tests your knowledge of the development of the three germ layers. The outer layer (ectoderm) develops into the nervous system, lens of the eye, as well as the skin. The middle layer (mesoderm) will develop into muscle and bone. The endoderm (inner layer) becomes the digestive tract.

53. (E)

This question about cellular metabolism tests your knowledge of the range of functions performed by the body on a chemical level. These include protein synthesis, cellular respiration, burning of fats, and conversion of sugars to fats. The transport of nutrients in the blood is not a metabolic function. These nutrients are passively carried by the plasma under the power of blood pressure.

54. (E)

The amount of growth is greater in the plant treated with gibberellic acid than in the control plant. However, this experiment does not measure the number of cells present or the size of the cells. Therefore, we cannot draw any conclusions about how growth is occurring.

55. (E)

DNA is a double stranded nucleic acid, with a sugar (deoxyribose) and phosphate backbone. RNA is single stranded with a sugar (ribose) and phosphate backbone. Both nucleic acids contain the bases adenine, guanine, and cytosine. In addition, DNA contains the base thymine, while RNA contains uracil.

56. (B)

This question asks you to distinguish between the cellular structure of plants and animals. Both types of cells contain mitochondria for ATP production, ribosomes for protein synthesis, Golgi bodies for packaging secretory products, and chromosomes for genetic information. Chloroplasts are only found in plants, and are the site of photosynthesis.

57. (A)

The reaction for photosynthesis is:

$$6\ CO_2 + 12\ H_2O + \text{light} \rightarrow C_6H_{12}O_6 + 6\ H_2O + 6\ O_2$$

(carbon dioxide) (glucose)

58. (E)

The primary purpose is to reduce the chromosome number from the normal diploid number in order to produce eggs and sperm with half that number (haploid). Then, when fertilization occurs the diploid chromosome number is restored. A side benefit of meiosis occurs during prophase I when like chromosomes pair up. At this point, crossing over can occur, and, thus, increase genetic variability. Meiosis in sperm production results in four sperm for each event, while meiosis in egg production gives only one egg. Because cell division is uneven in egg production, one of the cells receives most of the cytoplasm. The process of meiosis insures genetic variability, and, therefore, does not result in identical offspring.

Questions 59 - 62

This set of questions deals with the structure and function of organic molecules.

59. (C)

To answer this question correctly you need to know the basic components of organic compounds. The building blocks for carbohydrates are simple sugars, those for fats are glycerol and fatty acids, and those for proteins are amino acids. DNA contains phosphates, the sugar deoxyribose and four bases (adenine, cytosine, guanine, and thymine). RNA contains phosphates, the sugar ribose and four bases (adenine, cytosine, guanine, and uracil).

60. (D)

For this question, you need to know that a gene is a sequence of DNA nucleotides that codes for a particular polypeptide. The exceptions to this are RNA viruses. Their genes are sequences of RNA nucleotides. Since the question asks about plants and animals, the correct answer for this question is (D).

61. (A)

Photosynthesis is the process that living plants use to combine carbon dioxide and water in the presence of light to form oxygen and glucose which is a carbohydrate.

62. (C)

This question tests your ability to classify an enzyme as a protein. It is important to recognize the protein nature of the enzyme, since its chemical composition accounts for its properties. This includes the ability to chemically fit and react with a substance, the ability to be inhibited or poisoned, as well as the ability to be denatured.

Questions 63 - 64

This set of questions tests your knowledge of ecological concepts.

63. (B)

An ecosystem includes the organisms in a particular area plus the physical environment. That group of interacting organisms is called a community. Individuals within this community belonging to the same species are a population.

64. (E)

A niche includes the place where an organism lives (habitat), what it eats, when it is active, when it mates, its nutritional and energy sources, its rate of metabolism and growth and its interactions with other ecosystem operations.

Questions 65 - 68

This set of questions asks you to analyze and categorize particular organism interactions.

65. (A)

A deer tick inhabits as well as feeds off the deer by sucking its blood. This is an example of parasitism.

66. (E)

Two species of frogs are in this case related enough to share the same ecological needs. This might be food sources, avoidance of predators, etc. Therefore, they are competing for a particular place in the ecosystem.

67. (B)

The relationship of algae and fungi in lichens is clearly a case of both organisms deriving benefit from the association, an example of mutualism. The fungi capture the water needed to support photosynthesis. And, the algae undergo photosynthesis producing organic materials needed by the algae and fungi.

68. (D)

This is an example of commensalism, an association where one organism benefits, while the other is neither helped nor harmed.

Questions 69 - 70

This set of questions concerns the way organisms reproduce.

69. (D)

Budding is an asexual outgrowth that results in a new individual. Budding occurs in invertebrates, for example in annelids, coelenterates, and sponges, as well as yeasts.

70. (A)

Binary fission is an asexual reproduction that results in two identical cells by an equal pinching in half of the original cell. This occurs in prokaryotes.

Questions 71 - 74

For this set of questions you must know about the major events in development. They occur in the following order: cleavage, blastula formation, implantation, gastrulation, and neurulation.

71. (E)

The cells on the dorsal surface of the ectoderm invaginate (migrate inwards) to form the neural tube. This process of neurulation begins the formation of the central nervous system.

72. (A)

The cells at one end of the hollow blastula migrate inward forming a two layered structure. The inner layer becomes the digestive tract during this process of gastrulation.

73. (B)

Fertilization of the egg by the sperm triggers a series of cell divisions by mitosis called cleavage.

74. (C)

Trophoblasts are the outer layer of cells of the blastocyst. Prior to implantation, they enlarge and adhere to the uterine wall. Then, they secrete lytic enzymes which eat away at the endometrium, allowing the blastocyst to implant in the uterine wall.

Questions 75 - 78

This set of questions tests your knowledge of protein synthesis.

75. (C)
Messenger RNA is the copy of the gene that codes for a particular protein. It attaches at the ribosome and directs protein synthesis. Thus, the ribosome is the site of protein synthesis.

76. (E)
The amino acids are transported to the ribosome and properly oriented at the ribosome by transfer RNA.

77. (B)
A mutation is a heritable change in the genetic material, and can occur by alteration of the base sequence in DNA.

78. (A)
Messenger RNA forms in the nucleus as a copy of the DNA base sequence. It travels to the cytoplasm, attaches to the ribosome, and directs the assembly of amino acids into proteins.

Questions 79 - 83

These questions refer to the diagram of the human respiratory system. To answer the questions correctly you must identify the structures and associate each of them with their correct functions.

79. (E)
Gas exchange occurs in the alveoli of the lungs. The structure labeled 6 is where the alveoli are located.

80. (C)

Voice production occurs in the larynx. The larynx is structure 3.

81. (E)

For this question you need to know how air is inspired. The ribs rise, and the diaphragm contracts and flattens out. This increases the size of the thoracic cavity, which causes a decrease in its pressure. Thus, air moves into the lungs to equalize the pressure. The structure labeled 7 is the diaphram.

82. (C)

For this question you need to know about cleansing mechanisms in the respiratory system. One of the most important is the trapping of debris by cilia and mucus in the trachea. The one-way movement of the cilia helps to clear the trachea of minute particles (debris) of inhaled dust entrapped in the mucus. Structure 4 is the trachea.

83. (C)

For this question you must know about the underlying structure of the parts of the respiratory system. The C-shaped cartilaginous rings in the trachea help to hold the trachea open at all times. The trachea is labeled 4.

Questions 84 - 86

This group of questions refers to experiments on competition between two species of flour beetles when grown together in flour.

84. (B)

You are asked to determine which portion of Graph I shows the greatest rate of increase of *Tribolium* adults. Days 0 to 25 show no increase. Days 50 to 100, 100 to 150, and 150 to 200 show minor fluctuations, but no overall increase in numbers of adults. The largest increase (from 10 to 120 adults) occurs from day 25 to day 50.

85. (D)

In order to answer this question you need to compare the difference in growth of both species when grown in pure flour (Graph I) vs growth in the same amount of flour with glass tubing added (Graph II). Graph II shows more *Tribolium* adults and more *Oryzaephilus* adults than Graph I. Therefore, the presence of glass tubing has resulted in an increase in growth of both species.

86. (E)

For this question, an interpretation of the results involves an understanding of competition among organisms. Since Graph II demonstrates a decrease in competition, and since the only variable is the glass tubing, the glass tubing must be responsible for these results. Glass tubing would not affect the amount of flour eaten or the temperature of the flour. Therefore, the tubing must provide the flour beetles with more than one place to live, and, thus, decrease competition.

Questions 87 - 90

This set of questions tests your general knowledge of the circulatory system as well as your ability to interpret data about the circulatory system that is presented in graph form.

87. (C)

The capillaries have the slowest rate of blood flow with a velocity of 0.05 cm/sec. This is consistent with the function of the capillaries to allow for exchanges of materials.

88. (C)

To answer this question you must realize that the exchanges of nutrients, gases, and waste products occurs across blood vessels that are especially adapted for this purpose. These adaptions are found in the capillaries. They are both thin walled and contain blood flowing at a slow rate.

89. (D)

The ability to interpret the data presented in the table will give the answer to this question. We cannot make a direct correlation of blood pressure with cross-sectional area. This is because blood pressure decreases as the distance from the heart increases, while cross-sectional area increases then decreases. A velocity first decreases, then increases, while blood pressure continues to decrease. A comparison of cross-sectional area and velocity indicates that as the cross-sectional area increases, the velocity decreases.

90. (E)

An understanding of the anatomy of the circulatory system will lead to the answer to this question. The blood that leaves the heart, leaves via the left ventricle and enters the aorta. This is why the blood pressure is the highest in the aorta.

Questions 91 - 92

These two questions refer to the diagram of the carbon cycle.

91. (A) (diagram with processes labeled)

In order to answer this question you must know what processes are represented by the arrows in the diagram, and that the arrow from free carbon dioxide to producers represents a build up of carbon compounds. This process is photosynthesis.

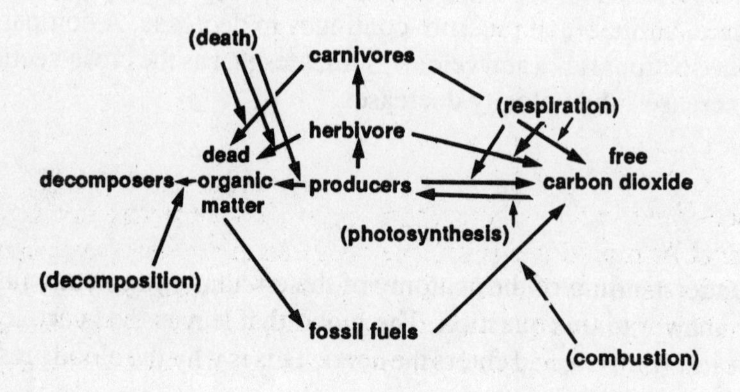

92. (B)

A decomposer breaks down dead organic matter and releases nutrients to the environment. Since the oak tree is a producer, it incorporates nutrients from the environment into organic matter. The caterpillar and horse are herbivores and use plants as their source of nutrients, while the lion is a carnivore and obtains nutrients by eating other animals. Only mushrooms live off dead organic matter and, therefore, are decomposers.

Questions 93 - 94

For this set of questions you must be familiar with experimental design and you must know that it is possible to induce mutations.

93. (C)
Petri plate I shows normal growth of the bacteria. Plate II containing minimal medium alone is the control for the experimental plates (plate III containing chemical A and plate IV containing chemical B). Thus, chemicals A and B are the variables in this experiment. Temperature is one of the conditions held constant during the experiment, and the colony counts are the results.

94. (D)
Since more colonies appear on plate III which was treated with chemical A, mutations must have occurred in the defective histidine gene. This gene now has the ability to produce histidine and can survive without histidine in the medium. Since plate IV containing chemical B has the same growth as the control (plate II), we must conclude that chemical B does not cause mutations of the defective histidine gene.

Questions 95 - 96

This set of questions refers to the hormonal control of metamorphosis in insects.

95. (C)
Since the transplant of a "young" corpus allatum causes a mature larva to continue to grow rather than pupate, the corpus allatum must behave differently depending upon its age. When it is present in an immature larva it is active, but it is no longer active in a mature larva. The production of juvenile hormone by the corpus allatum prevents ecdysone from causing pupation.

96. (B)
The prothoracic gland makes ecdysone which causes pupa formation or eclosion of adults. If this gland is removed, ecdysone will not be present, and the larva will continue to grow. We would not expect the larva to stop growing, since the corpus allatum does not regulate growth, but rather the ability of ecdysone to function.

Questions 97 - 100

This group of questions deals with the ability to interpret data from an experiment about osmosis.

97. (B)

When the eggs were exposed to different concentrations of salt, their diameters changed depending upon the concentration. The greatest change in diameter was 20% for eggs exposed to salt free water. Choice D is not correct, because water did move through the membrane but in equal amounts in both directions (see explanation for the next question).

98. (D)

In order to answer this question you mut know something about the relationship of animal cells to their environment. If the cells have a lower concentration of salt (hypotonic) in relationship to the environment, water will move out of the cell and the cell size will decrease. If the cells have a higher concentration of salt (are hypertonic) in relationship to the environment, water will move into the cell and increase their size. If the salt concentration is the same inside and out (isotonic), water will move in both directions equally and, thus, not change the overall size. Since there is no change in the diameter of the eggs exposed to 3.5% salt water, we would expect that the salt solution they are in matches the natural environment.

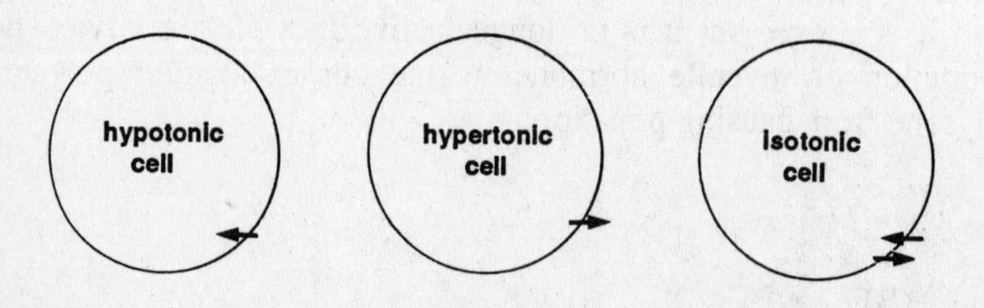

99. (C)

The membranes of living cells carefully control the internal salt concentration, but water moves according to its relative concentration. Because the inside of the cell has a particular salinity, water will move across the membrane from its area of higher concentration to its area of lower concentration, causing a change in cell diameter.

100. (E)

To answer this question you must know something about the structure of plants and the osmotic relationship of plants to their environment. Plants have a cell wall which prevents them from bursting no matter how hypotonic their environment. Plant cells in a hypertonic environment will lose water to the environment. If the environment is isotonic, there will be an equal exchange of water with the environment. When plant cells are in a hypotonic environment, they take in water and become firm. This is what is found naturally, and accounts for the turgor pressure characteristic of plant cells.

BIOLOGY

TEST II

BIOLOGY
ANSWER SHEET

1. (A) (B) (C) (D) (E)
2. (A) (B) (C) (D) (E)
3. (A) (B) (C) (D) (E)
4. (A) (B) (C) (D) (E)
5. (A) (B) (C) (D) (E)
6. (A) (B) (C) (D) (E)
7. (A) (B) (C) (D) (E)
8. (A) (B) (C) (D) (E)
9. (A) (B) (C) (D) (E)
10. (A) (B) (C) (D) (E)
11. (A) (B) (C) (D) (E)
12. (A) (B) (C) (D) (E)
13. (A) (B) (C) (D) (E)
14. (A) (B) (C) (D) (E)
15. (A) (B) (C) (D) (E)
16. (A) (B) (C) (D) (E)
17. (A) (B) (C) (D) (E)
18. (A) (B) (C) (D) (E)
19. (A) (B) (C) (D) (E)
20. (A) (B) (C) (D) (E)
21. (A) (B) (C) (D) (E)
22. (A) (B) (C) (D) (E)
23. (A) (B) (C) (D) (E)
24. (A) (B) (C) (D) (E)
25. (A) (B) (C) (D) (E)
26. (A) (B) (C) (D) (E)
27. (A) (B) (C) (D) (E)
28. (A) (B) (C) (D) (E)
29. (A) (B) (C) (D) (E)
30. (A) (B) (C) (D) (E)
31. (A) (B) (C) (D) (E)
32. (A) (B) (C) (D) (E)
33. (A) (B) (C) (D) (E)
34. (A) (B) (C) (D) (E)
35. (A) (B) (C) (D) (E)
36. (A) (B) (C) (D) (E)
37. (A) (B) (C) (D) (E)
38. (A) (B) (C) (D) (E)
39. (A) (B) (C) (D) (E)
40. (A) (B) (C) (D) (E)
41. (A) (B) (C) (D) (E)
42. (A) (B) (C) (D) (E)
43. (A) (B) (C) (D) (E)
44. (A) (B) (C) (D) (E)
45. (A) (B) (C) (D) (E)
46. (A) (B) (C) (D) (E)
47. (A) (B) (C) (D) (E)
48. (A) (B) (C) (D) (E)
49. (A) (B) (C) (D) (E)
50. (A) (B) (C) (D) (E)
51. (A) (B) (C) (D) (E)
52. (A) (B) (C) (D) (E)
53. (A) (B) (C) (D) (E)
54. (A) (B) (C) (D) (E)
55. (A) (B) (C) (D) (E)
56. (A) (B) (C) (D) (E)
57. (A) (B) (C) (D) (E)
58. (A) (B) (C) (D) (E)
59. (A) (B) (C) (D) (E)
60. (A) (B) (C) (D) (E)
61. (A) (B) (C) (D) (E)
62. (A) (B) (C) (D) (E)
63. (A) (B) (C) (D) (E)
64. (A) (B) (C) (D) (E)
65. (A) (B) (C) (D) (E)
66. (A) (B) (C) (D) (E)
67. (A) (B) (C) (D) (E)
68. (A) (B) (C) (D) (E)
69. (A) (B) (C) (D) (E)
70. (A) (B) (C) (D) (E)
71. (A) (B) (C) (D) (E)
72. (A) (B) (C) (D) (E)
73. (A) (B) (C) (D) (E)
74. (A) (B) (C) (D) (E)
75. (A) (B) (C) (D) (E)
76. (A) (B) (C) (D) (E)
77. (A) (B) (C) (D) (E)
78. (A) (B) (C) (D) (E)
79. (A) (B) (C) (D) (E)
80. (A) (B) (C) (D) (E)
81. (A) (B) (C) (D) (E)
82. (A) (B) (C) (D) (E)
83. (A) (B) (C) (D) (E)
84. (A) (B) (C) (D) (E)
85. (A) (B) (C) (D) (E)
86. (A) (B) (C) (D) (E)
87. (A) (B) (C) (D) (E)
88. (A) (B) (C) (D) (E)
89. (A) (B) (C) (D) (E)
90. (A) (B) (C) (D) (E)
91. (A) (B) (C) (D) (E)
92. (A) (B) (C) (D) (E)
93. (A) (B) (C) (D) (E)
94. (A) (B) (C) (D) (E)
95. (A) (B) (C) (D) (E)
96. (A) (B) (C) (D) (E)
97. (A) (B) (C) (D) (E)
98. (A) (B) (C) (D) (E)
99. (A) (B) (C) (D) (E)
100. (A) (B) (C) (D) (E)

BIOLOGY EXAM II

PART A

Directions: For each of the following questions or incomplete sentences, there are five choices. Choose the answer which is most correct. Darken the corresponding space on your answer sheet.

Question 1 refers to the diagram below.

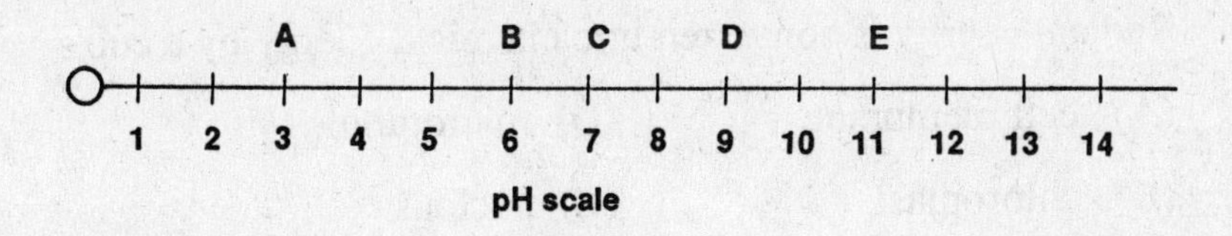

1. The pH closest to the normal pH of human skin surface is at:

(A) A
(B) B
(C) C
(D) D
(E) E

2. Which of the following elements is present in the smallest amount in cellular protoplasm?

(A) carbon
(B) hydrogen
(C) nitrogen
(D) oxygen
(E) phosphorus

3. The smallest type of cell in humans is a (n):

(A) nerve cell
(B) ovum
(C) skeletal muscle fiber
(D) red blood cell
(E) white blood cell

4. Radiant energy is converted into chemical energy by a cell's

(A) cell membrane
(B) chloroplast
(C) flagellum
(D) microtubule
(E) nucleus

5. In the genetic cross AaBb x aabb, assume that each lettered gene pair independently controls a separate trait of phenotype. The probability of an offspring appearing dominant for each trait is

(A) 0%
(B) 25%
(C) 50%
(D) 75%
(E) 100%

6. In the genetic cross that is mentioned in question 5, the probability of the occurrence of an offspring resembling neither parent is

(A) 0%

(B) 25%

(C) 50%

(D) 75%

(E) 100%

7. A protozoan cell is immersed in pure water. Select the correct statement from the following about the cell's behavior and environment.

(A) The cell will gain water.

(B) The cell will lose water.

(C) The exterior environment is hypertonic.

(D) The cell's interior is hypotonic

(E) The cell is isotonic to the outside.

Questions 8 and 9 refer to the drawing below.

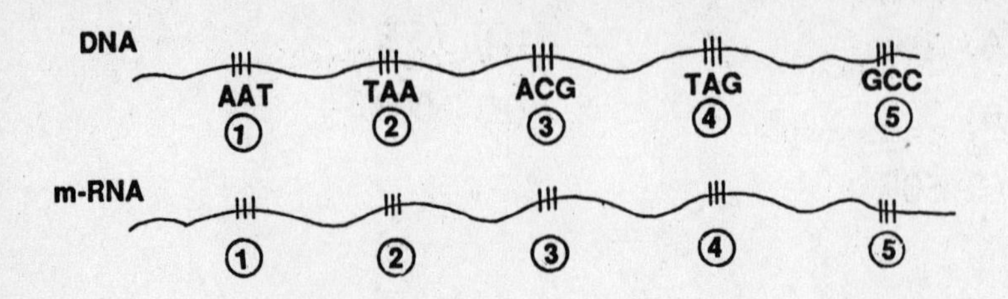

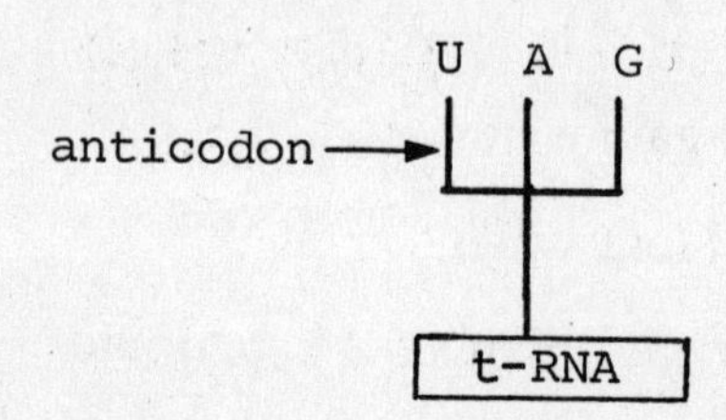

8. The third codon of the mRNA molecule is:

(A) ACG
(B) AGC
(C) TAG
(D) TGC
(E) UGC

9. The tRNA with its anticodon will couple with mRNA base triplet number:

(A) 1
(B) 2
(C) 3
(D) 4
(E) 5

10. White blood cells can engulf large foreign debris in the body by:

(A) active transport
(B) diffusion
(C) exocytosis
(D) phagocytosis
(E) pinocytosis

11. Cells can move particles of matter from areas of lower concentration to areas of higher concentration through the expenditure of energy. This process is termed:

(A) active transport
(B) bulk flow
(C) diffusion
(D) hydrolysis
(E) osmosis

12. Turgor is exhibited by a cell from a(n):

(A) alga
(B) *Amoeba*
(C) human
(D) *Paramecium*
(E) virus

13. A cell membrane consists primarily of molecules of:

(A) carbohydrates and lipids
(B) carbohydrates and proteins
(C) nucleic acids and proteins
(D) nucleic acids and phospholipids
(E) phospholipids and proteins

14. All of the following are needed components of the photosynthetic process except:

(A) carbon dioxide
(B) chlorophyll
(C) light
(D) oxygen
(E) water

15. Nutrient molecules usually produce two major forms of energy in an organism's body. They are:

(A) chemical and heat
(B) chemical and nuclear
(C) chemical and solar
(D) heat and nuclear
(E) nuclear and solar

16. Cells usually convert carbohydrate molecules into energy-currency molecules of:

(A) ADP
(B) ATP
(C) glucose
(D) oxygen
(E) starch

17. Glycolysis can best be described as:

(A) aerobic
(B) anaerobic
(C) citric
(D) lactic
(E) pyruvic

Questions 18 - 20 refer to the following diagram.

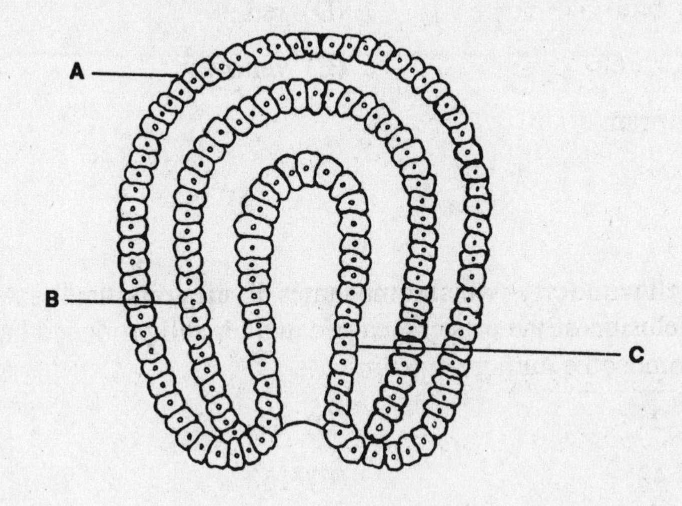

18. Embryonic layer A gives rise to:

(A) bones
(B) digestive tract
(C) kidney
(D) muscles
(E) nervous system

19. Embryonic layer B gives rise to:

(A) digestive tract
(B) kidney
(C) liver
(D) muscles
(E) skin

20. Embryonic layer C gives rise to:

(A) bones
(B) dermis
(C) digestive tract
(D) epidermis
(E) muscles

21. Select the light with the longest wavelength that is also absorbed during photosynthesis:

(A) blue
(B) green
(C) orange
(D) red
(E) yellow

22. A cell with forty-two chromosomes divides mitotically. After conclusion of the process, each daughter cell produced has a chromosome number equaling:

(A) 21
(B) 42
(C) 44
(D) 84
(E) 168

23. Mitosis produces cells that are:

(A) diploid
(B) haploid
(C) homologous
(D) tetraploid
(E) zygotes

24. Meiosis produces cells that are:

(A) diploid
(B) haploid
(C) homologous
(D) tetraploid
(E) zygotes

25. A parent cell with 64 chromosomes engages in meiosis. At the end of the second division process, each daughter cell has a chromosome number equaling:

(A) 8
(B) 16
(C) 32
(D) 64
(E) 128

26. Another name for throat is:

(A) esophagus
(B) larynx
(C) nasal cavity
(D) pharynx
(E) trachea

27. Which of the following organelles undergoes division during mitosis?

(A) ribosome
(B) lysosome
(C) nucleolus
(D) mitochondrion
(E) none of the above

28. A genetic counselor offers two prospective parents a 100% chance for producing a child who will develop diabetes mellitus. The genotype of the two parents is:

(A) both heterozygous
(B) both homozygous dominant
(C) both homozygous recessive
(D) one homozygous dominant, one heterozygous
(E) one homozygous recessive, one heterozygous

29. A DNA strand has a base sequence of CGTAGT. The messenger RNA transcribed from it has a base sequence of:

(A) ACUACG
(B) CGTAGT
(C) GCAUCA
(D) GACUAC
(E) UACGUA

30. The protein binding and blocking the operator gene in the operon model of gene control is the:

(A) inactivator
(B) operation
(C) promoter
(D) regulator
(E) repressor

31. After storage in the epididymis, migrating sperm cells next encounter the:

(A) penis
(B) prostate
(C) testis
(D) urethra
(E) vas deferens

32. A messenger RNA molecule with three hundred bases contains the codes for how many amino acids?

(A) 3
(B) 100
(C) 300
(D) 600
(E) 900

33. Pathogenic bacteria:

(A) are autotrophs

(B) are beneficial to humans

(C) cause disease

(D) die easily

(E) live constantly in the soil

34. Which of the following organisms is ciliated?

(A) *Amoeba*

(B) *Clostridium*

(C) *Euglena*

(D) *Paramecium*

(E) *Trypanosoma*

Questions 35 and 36 refer to the diagram below.

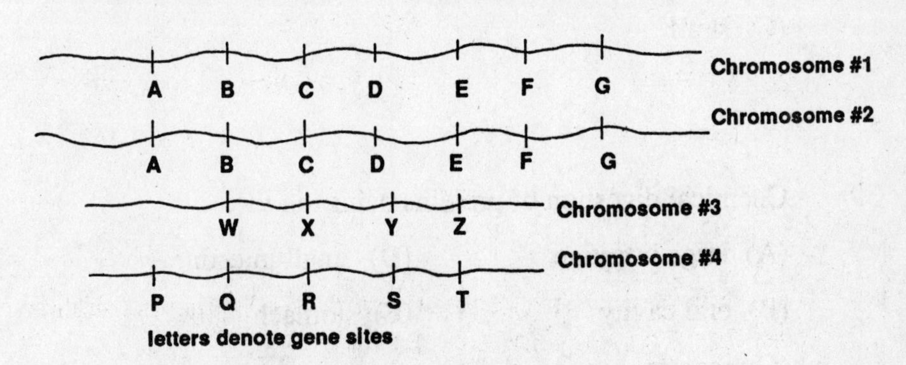

35. A deletion could yield chromosome:

(A) ABCDEFGH

(B) ABCDEFGEFG

(C) PQRST

(D) WXY

(E) WXYZPQ

36. A translocation could yield chromosome:

(A) ABCDEFG
(B) ABCABCDEFG
(C) ABCWXYZ
(D) WXYZ
(E) PQRS

37. Which plant tissue matures to become wood in an aging tree?

(A) epidermis
(B) collenchyma
(C) cortex
(D) phloem
(E) xylem

38. Select the bone that does not belong to the axial skeleton.

(A) humerus
(B) rib
(C) skull
(D) sternum
(E) vertebra

39. Chemical digestion of proteins beings in the:

(A) large intestine
(B) oral cavity
(C) rectum
(D) small intestine
(E) stomach

40. Skeletal muscle function is best summarized by the word:

(A) appearance
(B) conduction
(C) contraction
(D) protection
(E) shaping

41. Impulses normally traverse the regions of a neuron in which of the following orders?

(A) axon, cell body, dendrite
(B) axon, dendrite, cell body
(C) cell body, dendrite, axon
(D) dendrite, axon, cell body
(E) dendrite, cell body, axon

42. The portion of the brain continuous with the spinal cord is the:

(A) cerebellum
(B) cerebrum
(C) medulla
(D) pons
(E) thalamus

43. Which endocrine gland secretes human growth hormone?

(A) adrenal
(B) pancreas
(C) parathyroid
(D) pituitary
(E) thyroid

Questions 44 - 45 refer to the phylogenetic tree of animal phyla below.

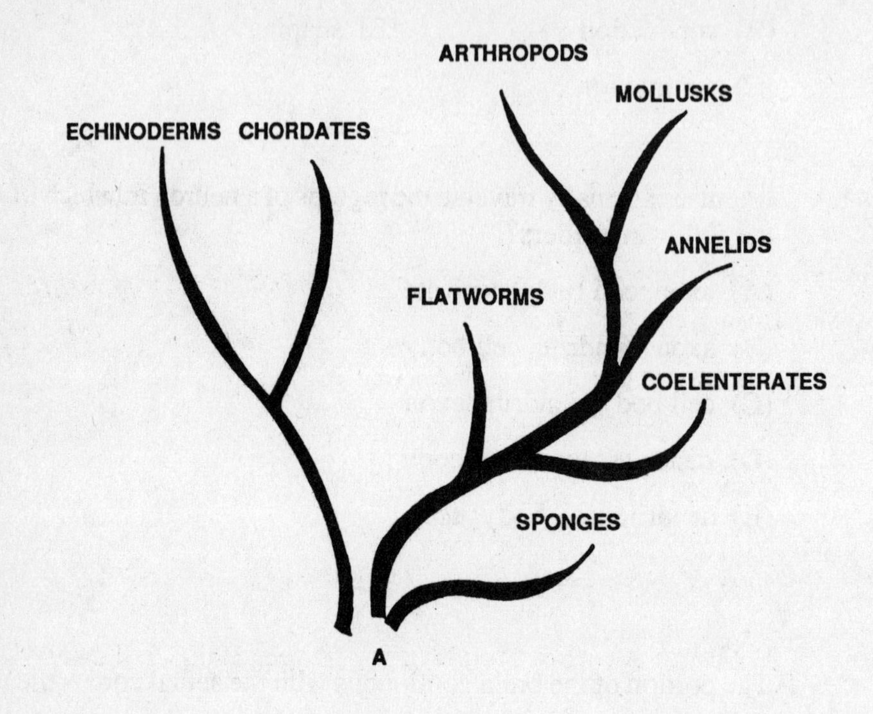

44. A was:

(A) a plant

(B) an alga

(C) an arthropod

(D) one-celled

(E) a vertebrate

45. Which two phyla are the most distantly related?

(A) annelids and arthropods

(B) arthropods and mollusks

(C) coelenterates and flatworms

(D) echinoderms and chordates

(E) echinoderms and mollusks

46. The most numerous and microscopic type of air-flow passage is the:

(A) alveolus
(B) bronchiole
(C) bronchus
(D) larynx
(E) trachea

47. Nutrient absorption mainly occurs in what portion of the gastrointestinal tract?

(A) large intestine
(B) small intestine
(C) pharynx
(D) rectum
(E) stomach

48. The parathyroid hormone regulates concentration in the blood of:

(A) calcium
(B) chloride
(C) magnesium
(D) potassium
(E) sodium

49. A correct summary of the normal human male chromosome complement is:

(A) 42 autosomes, 2 X chromosomes, 2 Y chromosomes
(B) 44 autosomes, 2 X chromosomes
(C) 44 autosomes, 1 X chromosome, 1 Y chromosome
(D) 46 autosomes, 2 X chromosomes
(E) 46 autosomes, 1 X chromosome, 1 Y chromosome

50. Which characteristic differentiates RNA from DNA? RNA:

(A) contains the base uracil

(B) has a phosphate group

(C) has five bases

(D) is double-stranded

(E) lacks a sugar in its nucleotide

51. A source of genetic change in a population is:

(A) catastrophism

(B) fossils

(C) gene flow

(D) mutations

(E) natural selection

52. A female carrier of hemophilia marries a male hemophiliac. What percent of their daughters can they expect to develop hemophilia?

(A) 0

(B) 25

(C) 50

(D) 75

(E) 100

53. Which of the following is a pioneer organism?

(A) man

(B) tree

(C) insect

(D) lichen

(E) fungus

54. The number of adenine residues in a DNA double helix equals the number of residues of

(A) adenine
(B) cytosine
(C) guanine
(D) thymine
(E) uracil

55. A community of organisms interacting with abiotic environ - mental factors composes a(n):

(A) biogeochemical cycle
(B) food chain
(C) ecosystem
(D) niche
(E) population

56. How many different phenotypes can be produced in a dihybrid cross?

(A) 2
(B) 4
(C) 6
(D) 8
(E) 10

57. Forty-five autosomes and two sex chromosomes are observed in the karyotype of a human afflicted with which of the following syndromes?

(A) Down's
(B) Klinefelter's
(C) Mendel's
(D) Turner's
(E) Watson's

58. The kidneys pass urine to the next part of the urinary tract, which is the:

(A) bladder
(B) calyx
(C) pelvis
(D) ureter
(E) urethra

59. Select the female reproductive structure of a plant.

(A) anther
(B) filament
(C) phloem
(D) stamen
(E) style

60. In twenty-four hours, a pair of human kidneys excretes 2 liters of urine while filtering 150 liters of extracellular fluid. The amount they reabsorb is:

(A) 75
(B) 148
(C) 150
(D) 152
(E) 300

61. Human sex cell fertilization most often occurs in the:

(A) ovary
(B) oviduct
(C) testis
(D) uterus
(E) vagina

62. Which naturalist independently came to a conclusion similar to Darwin's natural selection principle?

(A) Cuvier
(B) Lamarck
(C) Lyell
(D) Malthus
(E) Wallace

63. For a given trait, a randomly-mating population has established a dominant gene frequency of 70% in its gene pool. The frequency of homozygous recessive individuals in the population is:

(A) .03
(B) .09
(C) .27
(D) .3
(E) .7

PART B

Directions: Each of the following sets of questions refers to a list of lettered choices immediately preceding the question set. Choose the best answer from the list, then darken the corresponding space on your answer sheet. Choices may be used once, more than once, or not at all.

Questions 64 - 67

(A) carbohydrates
(B) lipids
(C) nucleic acids
(D) proteins
(E) steroids

64. store the most calories per gram; serve in body insulation.

65. essence of genetic structure

66. principal energy-storing molecules, such as glucose and starch

67. structural molecules formed by amino acids; part of enzyme structures

Questions 68 - 70

(A) alga
(B) bass
(C) human
(D) minnow
(E) sunfish

Assume that all of the organisms that are listed above are members of a single food chain.

68. Which organism is a primary consumer?

69. The least amount of useful energy is available to which organism?

70. Which organism is the producer?

Questions 71 - 74

(A) arteriole
(B) artery
(C) capillary
(D) vein
(E) venule

71. supplier of blood to microscopic blood vessels

72. has a higher blood pressure than any other type of vessel

73. returns blood directly to the heart

74. site of exchange of substances with body cells

Questions 75 - 76

(A) bullfrog
(B) rattlesnake
(C) robin
(D) shark
(E) trout

75. The vertebrate class Chondrichthyes contains the member:

76. The vertebrate class Osteichthyes contains the member:

Questions 77 - 80

(A) amensalism
(B) commensalism
(C) mutualism
(D) parasitism
(E) predation

77. the consumption of other organisms

78. a relationship in which one organism lives at the expense of another

79. a relationship in which both organisms benefit from the association

80. a relationship in which one organism benefits, while another organism is unaffected

PART C

Directions: The following questions refer to the accompanying diagram. Certain parts of the diagram have been labelled with numbers. Each question has five choices. Choose the best answer and darken the corresponding space on your answer sheet.

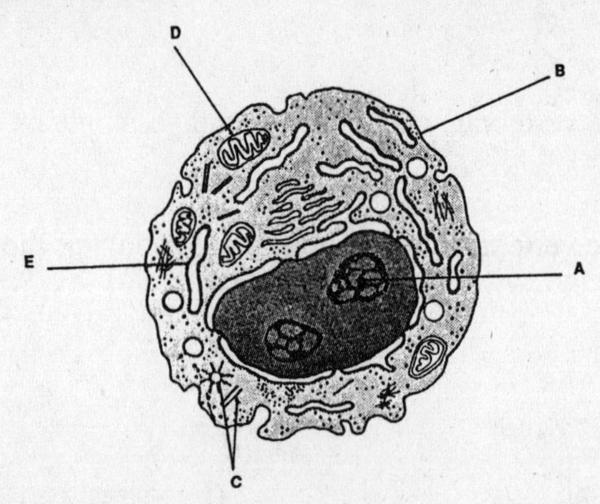

81. Large amounts of RNA are synthesized in the structure that is labelled:

(A) A
(B) B
(C) C
(D) D
(E) E

82. Internal transport is possible in this cell due to the structure that is labelled:

(A) A
(B) B
(C) C
(D) D
(E) E

83. Surrounded by astral rays is the structure that is labelled:

(A) A
(B) B
(C) C
(D) D
(E) E

84. ATP molecules are generated by the structure that is labelled:

(A) A
(B) B
(C) C
(D) D
(E) E

PART D

Directions: Each of the following questions refer to a laboratory or experimental situation. Read the description of each situation. Then select the best answer to each question. Darken the corresponding space on the answer sheet.

Laboratory Data - Enzyme Activity

Temperature (Centigrade)	Time Required (min.) to Convert 1 gram of Substance
0	25
5	20
20	15
37	5
70	25

85. Select the correct statement summarizing the relationship between temperature and enzyme activity in this experiment.

(A) As temperature decreases, enzyme activity increases

(B) As temperature increases, enzyme activity increases

(C) Enzyme activity is least at 37 degrees

(D) Enzyme activity peaks at 37 degrees but is decreased at temperature extremes

(E) Enzyme activity peaks at 70 degrees

Questions 86 and 87 refer to the cross section of a leaf viewed through a microscope.

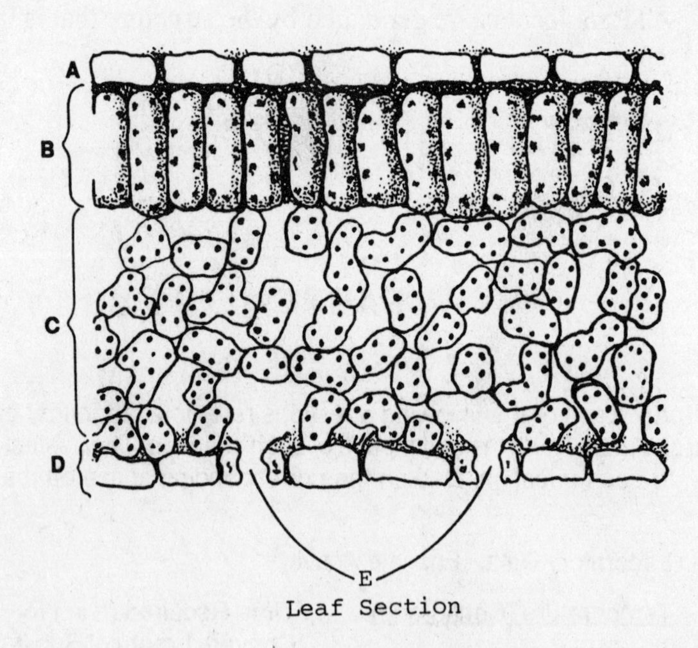

Leaf Section

86. The epidermal layer without stomata is labelled:

(A) A
(B) B
(C) C
(D) D
(E) E

87. An accurate statement about the layer of cells that is labelled B is:

(A) 100% humidity exists between cells here
(B) photosynthesis can occur rapidly here
(C) stomata work actively here
(D) this is the spongy mesophyll
(E) this layer contains the leaf cuticle

Questions 88 and 89 refer to the experimental findings required to grow a crop of corn. Results are summarized in the table below.

Nutrient	Pounds required to grow 100 bushels of corn per acre
N	160
K	80
P	40
Zn	3
Mg	.5
Mn	.2
Cu	.1
Mb	.05

88. The element to be incorporated into plant protein structure is:

(A) K
(B) Mg
(C) Mn
(D) N
(E) Zn

89. Which of the following elements is a macronutrient?

(A) copper
(B) magnesium
(C) manganese
(D) potassium
(E) zinc

Questions 90 - 92 refer to the table below. Solve in each case for the unsolved air volume from 3 human subjects tested in the lab:

Respiratory Rate	Tidal Volume (ml)	Respiratory Minute Volume (ml)	Expiratory Reserve Volume (ml)	Inspiratory Reserve Volume (ml	Vital Capacity (ml)
12	500	6000	(A)	2500	5000
10	400	4000	1500	2400	(B)
15	(C)	9000	2000	3000	5000

These values were recorded from three separate subject readings during an experiment concerning spiromatry – the measurement of air-breathing volumes.

These values were recorded from three separate subject readings during an experiment concerning spirometry - the measurement of air-breathing volumes.

90. Unrecorded volume A is:

(A) 1000
(B) 2000
(C) 3000
(D) 4000
(E) 5000

91. Unrecorded volume B is:

(A) 3900
(B) 4300
(C) 4400
(D) 5500
(E) 8300

92. Unrecorded volume C is:

(A) 100
(B) 200
(C) 300
(D) 600
(E) 1000

Questions 93 and 94 refer to the graph below.

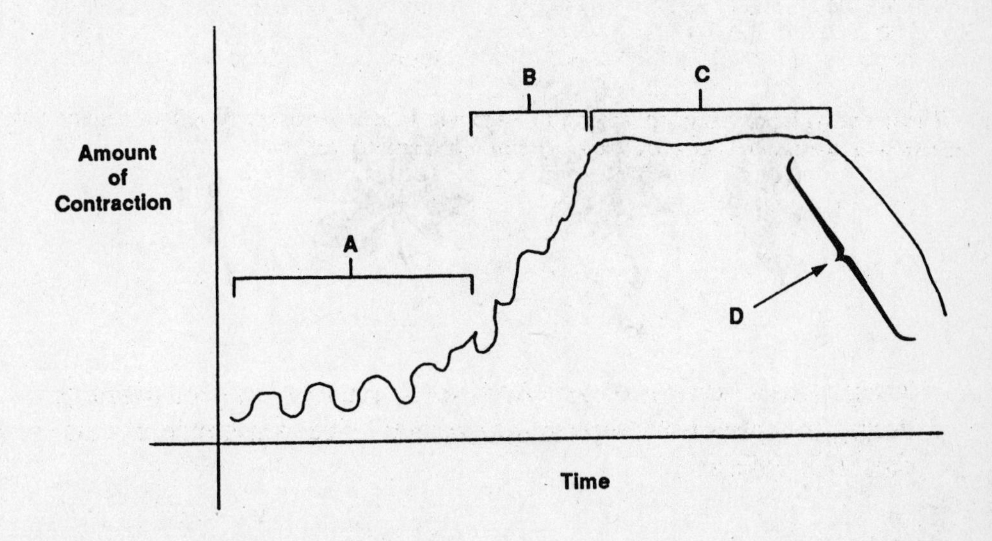

93. The graph depicts all patterns of muscle contraction except:

(A) fatigue
(B) tetanus
(C) tonus
(D) simple twitch
(E) summation

94. The muscle reponse in region D is due to:

(A) accumulation of lactic acid
(B) excess glucose
(C) lack of stimulation
(D) presence of ATP
(E) wasting of muscle proteins

Questions 95 - 98 refer to the plant leaf drawings and a short dichotomous key for their classification below. Work each plant leaf through the statement pairs to arrive at each plant's genus name.

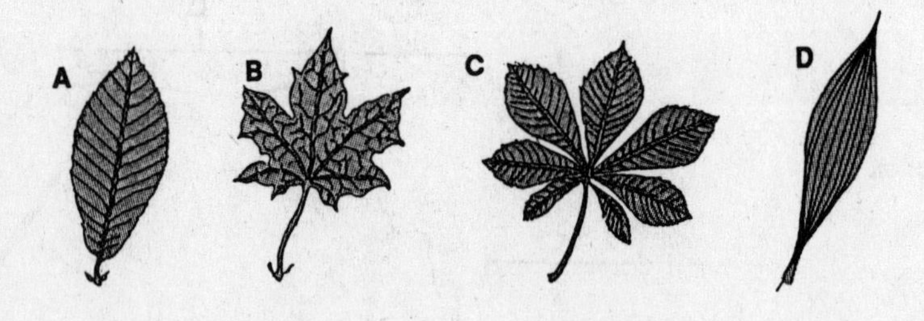

Key

1. (A) leaf has one terminal point - go to no. 2

   (B) leaf has more than one terminal point - go to no. 3

2. (A) leaf has parallel venation - *Rubrus*

   (B) leaf has pinnate venation - *Copernicus*

3. (A) leaf has palmate venation - *Acer*

   (B) leaf has palmately compound venation - *Aesculus*

4. leaf has none of these characteristics - *Libra*

95. Organism A belongs to the genus:

    (A) *Acer*
    (B) *Copernicus*
    (C) *Libra*
    (D) *Rubrus*
    (E) *Aesculus*

96. Organism B belongs to the genus:

    (A) *Acer*
    (B) *Coperincus*
    (C) *Libra*
    (D) *Rubrus*
    (E) *Aesculus*

97. Organism C belongs to the genus:

    (A) *Acer*
    (B) *Copernicus*
    (C) *Libra*
    (D) *Rubrus*
    (E) *Aesculus*

98. Organism D belongs to the genus:

(A) *Acer*

(B) *Copernicus*

(C) *Libra*

(D) *Rubrus*

(E) *Aesculus*

Questions 99 - 100 refer to the drawing below.

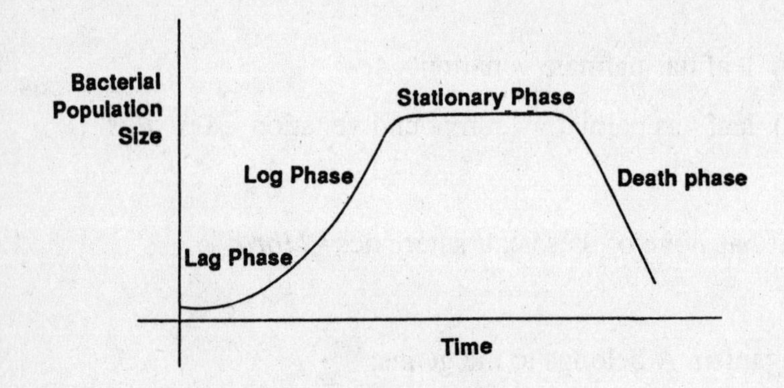

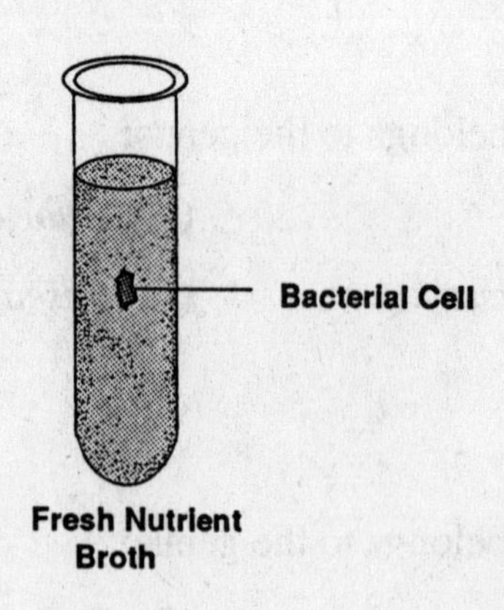

99. One bacterial cell is placed into a nutrient broth in a test tube at noon. Its generation time is 20 minutes. By 2:00 p.m., the size of the population of bacteria in the test tube is:

(A) 2
(B) 16
(C) 32
(D) 64
(E) 128

100. A correct statement about the population's stationary phase is:

(A) cells are not dying
(B) cells are not reproducing
(C) cell birth rate equals cell death rate
(D) cell death rate excedes cell birth rate
(E) cells reproduce too rapidly

BIOLOGY TEST II

ANSWER KEY

1.	B	34.	D	67.	D
2.	E	35.	D	68.	D
3.	D	36.	C	69.	C
4.	B	37.	E	70.	A
5.	B	38.	A	71.	A
6.	C	39.	E	72.	B
7.	A	40.	C	73.	D
8.	E	41.	E	74.	C
9.	D	42.	C	75.	D
10.	D	43.	D	76.	E
11.	A	44.	D	77.	E
12.	A	45.	E	78.	D
13.	E	46.	A	79.	C
14.	D	47.	B	80.	B
15.	A	48.	A	81.	A
16.	B	49.	C	82.	E
17.	B	50.	A	83.	C
18.	E	51.	D	84.	D
19.	D	52.	C	85.	D
20.	C	53.	D	86.	A
21.	D	54.	D	87.	B
22.	B	55.	C	88.	D
23.	A	56.	B	89.	D
24.	B	57.	A	90.	B
25.	C	58.	D	91.	B
26.	D	59.	E	92.	D
27.	D	60.	B	93.	C
28.	C	61.	B	94.	A
29.	C	62.	E	95.	B
30.	E	63.	B	96.	A
31.	E	64.	B	97.	E
32.	B	65.	C	98.	D
33.	C	66.	A	99.	D
				100.	C

BIOLOGY TEST II

DETAILED EXPLANATIONS OF ANSWERS

SECTION I

1. (B)

The pH of human skin is about 5.5 due to the secretion of fatty acids by sebaceous glands in the skin.

2. (E)

Oxygen contributes the highest percentage of protoplasm's weight. Carbon, hydrogen and nitrogen follow consecutively in their abundance. The percentage of weight of protoplasm contributed by phosphorus is much less than that of any of the others, usually less than 1%.

3. (D)

Red blood cells are the smallest body cells, seven to eight micrometers in diameter. White blood cells are slightly larger. Nerve and skeletal muscle cells are larger, the human ovum is large enough to be visible without using a microscope, approximately 1.5 mm in diameter.

4. (B)

Chloroplasts contain chlorophyll molecules, which, when irradiated by light of the proper wavelength, pass certain of their electrons (which have been raised to a higher, less stable energy level due to their absorption of light energy) to an electron transport chain. The reactions that occur along this chain allow the energy of the original excited chlorophyll electrons to be stored temporarily in reduced NADP and ATP.

5. (B)

Consider the following Punnett square that summarizes this genetic cross. On the far left block of the genetic grid, there is a genotype AaBb. This is the only genotype endowing the organism with at least one dominant allele in each gene pair for development of the dominant phenotype of each trait. The other three genotypes, Aabb, aaBb, and aabb, do not.

	AB	Ab	aB	ab
ab	AaBb	Aabb	aaBb	aabb

6. (C)

Two genotypes, AaBb and aabb, resemble one or the other parent. The other two, Aabb and aaBb, do not.

7. (A)

By osmosis, water will flow through the microorganism's cell membrane from the outside area of higher (100%) water concentration to inside the cell, where the concentration is lower. Choices C and D describe the reverse of the true situation. The cell's exterior is hypotonic (has a lower solute concentration and a higher water concentration) whereas the cell's interior is hypertonic (has higher solute levels).

8. (E)

A codon is a base triplet. Its identity is determined by the corresponding DNA codon. The third DNA codon, ACG, transcribes UGC in a manner compliant with the following complementary base-pairing rules:

DNA	RNA
A.	U
C.	G
G.	C
T.	A

9. (D)

The tRNA has an anticodon of UAG, which is complementary to the mRNA codon AUC. This must be mRNA codon #4, as the fourth DNA codon, TAG, determines the mRNA codon by transcription. RNA base-pairing rules are:

mRNA	tRNA
A.	U
C.	G
C.	C
U.	A

10. (D)

Phagocytosis is "cellular eating," which a white blood cell performs as part of the human body's defense system. Pinocytosis is "cellular drinking." Exocytosis means the removal of substances from a cell. The other two choices are unrelated transport processes of cell molecules, atoms, or ions.

11. (A)

The statement is a rote definition of active transport, which acts in opposition to diffusion and osmosis. The other choices are unrelated to energy requiring transport processes.

12. (A)

Plant or plantlike cells can take up water osmotically, yet not burst due to their rigid outer cell walls. The tremendous incompressible fluid buildup causes pressure buildup, or turgor. Choices (B) through (E) are not plants or plantlike.

13. (E)

This choice states the simple fact that the two most abundant molecules in cell membranes are phospholipids and proteins. The other listed molecules are not as abundant in the membrane.

14. (D)

Atoms of carbon dioxide and water molecules are used to <u>form</u> sugar and oxygen molecules. Chlorophyll molecules trap sunlight energy for use in the production of sugars. Oxygen is <u>produced</u>, not used, during photosynthesis.

15. (A)

Energy from the food you eat is stored chemically in certain molecular forms (glycogens, lipids) or is converted to heat, which maintains body temperature. Types of energies in the other choices are not appropriate for body use.

16. (B)

ATP, adenosine triphosphate, is the universal "energy currency" of cells. Metabolic pathways of glycolysis, the Krebs cycle, and pathways leading to them, convert molecules not immediately usable into a form that is "spendable."

17. (B)

Anaerobic means "without oxygen;" and glycolysis is an oxygen-independent series of reactions leading to the formation of a molecule, pyruvic acid. Pyruvic acid can then enter the oxygen-dependent (aerobic) citric acid cycle (Krebs cycle).

18. (E)

The outer early embryonic layer is the ectoderm. Its cells differentiate to give rise to the epidermis and the nervous system.

19. (D)

The middle layer is the mesoderm. It gives rise to bones, blood and muscles.

20. (C)

The inner layer is the endoderm. It gives rise to many internal organs such as those of the digestive tract.

21. (D)

Of all the types of light that are absorbed by chlorophyll, red light has the greatest wavelength (700 nm).

22. (B)

Barring a rare mutation, mitosis preserves the constancy of a cell's (and its organism's) genetic structure. This includes the somatic cell chromosome number for that species.

23. (A)

Mitosis produces daughter cells whose chromosome numbers and gene content are the same as those of the parent cell. A cell is referred to as being diploid if it contains two of each type of chromosome. For example, the 46 human chromosomes are maintained as 23 homologous pairs.

24. (B)

Meiosis produces sex cells that have the haploid number of chromosomes. In the process, chromosome pairs are separated, so sex cells get single chromosome copies. A cell is haploid if it contains only one of each type of chromosome.

25. (C)

Meiosis halves the chromosome number. If sex cells fertilize, sex cells with the halved chromosome number recombine to form a zygote with the full chromosome complement.

26. (D)

The pharynx is a region in which the respiratory and digestive tracts cross over. Air moves through the nasal cavity (more anterior) into the pharynx and on to the larynx (ventral). The oral cavity (more posterior) passes food into the pharynx end of the esophagus (dorsal).

27. (D)

Mitochondria, which contain their own DNA, undergo division while the cells that they are in undergo mitosis. This means that each daughter cell will have approximately as many mitochondria as the parent cell had. Ribosomes and lysosomes do not undergo division, and nucleoli are not present during mitosis.

28. (C)

Diabetes mellitus is produced by a recessive gene on an autosome pair (any chromosome pair other than the sex-determining pair). Thus, a homozygous recessive genotype, aa, produces this hereditary disease. If this is to occur with a 100% chance, both parents must also be aa to ensure inheritance of the gene twice, without other possibilities. If, for example, one parent were Aa, the chances for a diabetic offspring are 50%, not 100%.

29. (C)

DNA-to-RNA complementarity depends on base-pairing rules, which are as follows:

DNA	RNA
A	U
C	G
G	C
T	A

The GCAUCA base-for-base readout is the only choice compatible to the given DNA. These base-pairing rules must be memorized.

30. (E)

The repressor is the protein made by an adjacent regulator gene, and it can bind to the operator. The adjacent promoter gene is the site at which the RNA polymerase binds to DNA. Inactivator and operation have nothing to do with the operon model of gene control.

31. (E)

Sperm travel through the following structures in the following order: testis, epididymis, vas deferens, prostate gland, and the urethra. The urethra passes sperm on to and through the penis.

32. (B)

Three mRNA bases (codon) constitute the code for one amino acid attached to its tRNA. Thus, 300 to 100 is the proper ratio.

33. (C)

Pathogenic means disease-producing. The other choices do not describe its meaning. The vast majority of bacteria are not pathogenic, but beneficial to mankind and the ecology of planet earth.

34. (D)

An *Amoeba* moves using pseudopodia. The *Euglena* and *Trypanosoma* employ flagella. *Clostridium* is a bacterium that has no cilia or flagella.

35. (D)

A deletion is the loss of a fragment from a chromosome. Chromosome #3, WXYZ, has lost gene Z to become WXY.

36. (C)

A translocation occurs when a fragment of a chromosome attaches to a nonhomologous chromosome. Genes ABC from chromosome #1 or #2, a homologous pair, have attached to nonhomologous chromosome WXYX to yield the new combination ABCWXYZ.

37. (E)

New xylem is a tissue that conducts water and dissolved minerals through a plant body. With age, however, it undergoes chemical changes and hardens into wood. It then no longer functions as a conductive tissue. Phloem conducts organic molecules and collenchyma is a supportive tissue. Epidermis covers the plant and the cortex is a tissue that stores materials.

38. (A)

The axial skeleton is the upright central axis of the skeleton: skull, vertebral column, and rib cage (12 pairs of ribs with sternum). The other branch of the skeleton is the appendicular: appendages and girdles (shoulder and hip) that attach them to the axis. The humerus is the upper arm bone, and is therefore part of an appendage.

39. (E)

Enzymes that facilitate the chemical breakdown of proteins are lacking in the oral cavity. The stomach is the first structure of the alimentary canal that secretes a proteinase, pepsin.

40. (C)

Muscles shorten their length, thus exacting a pulling force on bones to produce movement.

41. (E)

This series lists the neuronal regions from end to end, over which the impulse travels. Dendrites, short and branching, carry the impulse to the cell body, which houses the neuron's nucleus and the majority of its cytoplasm. The long, single axon is the other cell process stemming from the cell's opposite end.

42. (C)

The medulla is at the base of the brainstem. The pons is anterior to it and the forebrain's thalamus is even more anterior in position. The cerebrum is the most anterior portion of the forebrain, whereas the cerebellum is dorsal to the hindbrain (pons and medulla) and is not continuous with the spinal cord.

43. (D)

This is a straightforward fact although the other listed endocrine glands also have their hormonal secretions: adrenal - epinephrine; pancreas-insulin; parathyroid-parathyroid horomone; and thyroid - thyroid hormone.

44. (D)

A is at the base of this evolutionary tree, the suggested one-celled forerunner of all evolving multicellular animal phyla.

45. (E)

The echinoderms and mollusks are from separate, widely-divergent evolutionary lines. The other four choices offer phylum pairs of more related phyla.

46. (A)

A pair of human lungs is estimated to have 300 million alveoli exchanging oxygen and carbon dioxide with blood flowing to and from the heart. Air inhaled through the tract travels through the following structures: nasal cavity, pharynx, larynx, trachea, primary bronchus, secondary bronchus, bronchiole, alveolar duct, and alveolus.

47. (B)

Chemical digestion occurs in the oral cavity, stomach, and small intestine. Digested subunits - amino acids, monosaccharides, and fatty acids - then pass from the tract into the blood stream.

48. (A)

Parathyroid hormone raises calcium levels in the blood by pulling calcium out of the bones, enhancing its reabsorption in the kidney and absorption in the small intestine.

49. (C)

An autosome is any chromosome in the complement, excepting the sex-determining pair - XX in females or XY in males. In a normal male somatic cell complement of 46 chromosomes, there are 44 (22 pairs) of autosomes, one X-chromosome, and one Y-chromosome.

50. (A)

Both nucleic acids have adenine, cytosine, or guanine as three of their four possible nucleotide bases. However, their fourth bases differ - - DNA contains thymine, and RNA contains uracil. Both have nucleotide phosphate groups. The other statements about RNA are untrue.

51. (D)

A population cannot change without a source of change. If only one form of a gene existed for a trait, there would not be variation. All organisms would be homozygous for that gene. Their offspring would be the same. Mutations are one way in which a population can obtain genetic variability.

52. (C)

Hemophilia is an X-linked recessive trait; therefore, the couple's offspring can be expected to have the following genotypes:

	X	X_H
X_H	XX_H	X_HX_H
Y	XY	X_HY

Thus, 50% of the couple's daughters will develop hemophilia.

53. (D)

A pioneer organism is the first organism to populate a region. It must be autotrophic and able to withstand the harshest conditions. Lichens require little water and can survive a variety of temperatures. Lichens can be found on bare rock and in the Antarctic.

54. (D)

Thymine is the choice because of the A-T base-pairing rule for complementarity. Every thymine base will be coupled with an adenine on the corresponding opposite half of the double helix.

55. (C)

This is a straight definition of an ecological system.

56. (B)

Study of this other historically famous Mendel cross reveals probabilities of: 9/16 for phenotypes dominant for both studied traits, 3/16 dominant for one trait, 3/16 dominant for the other trait and a 1/16 chance for a phenotype that is a result of homozygous recessive genotype for both traits.

AaBb x AaBb

	AB	Ab	aB	ab
AB	AABB	AABb	AaBB	AaBb
Ab	AABb	AAbb	AaBb	Aabb
aB	AaBB	AaBb	aaBB	aaBb
ab	AaBb	Aabb	aaBb	aabb

57. (A)

A person with Down's syndrome has an extra chromosome 21, instead of the normal pair. Thus, instead of 22 autosome pairs (44) and a pair of sex chromosomes, a person afflicted with Down's syndrome has 42 autosomes, three chromosome 21, and a pair of sex chromosomes, all of which total 47. Turner's syndrome patients have one less chromosome than they do normally. Klinefelter's victims do have 47, but they have 44 autosomes and 3 sex chromosomes.

58. (D)

The ureter is a ten-to-twelve-inch tube that takes urine from the kidney and conducts it to the bladder. The urethra carries urine from the bladder to the exterior of the body. The pelvis and calyx are not part of the urinary tract (the renal pelvis is, however, part of the kidney).

59. (E)

Phloem is conductive tissue and the other three structures are male reproductive structures. The style is a slender stalk with an ovary at its base. A pollen tube from the pollen grain grows down through it.

60. (B)

Amount excreted equals amount filtered minus amount reabsorbed. Thus, 150 - 148 = 2.

61. (B)

A migrating sperm cell usually meets the ovum early after the egg's release from the ovary. Unusual (ectopic) sites of fertilization may occur but are rare. The sperm moves beyond the vagina and uterus for oviduct fertilization.

62. (E)

This is a fact. Wallace seldom receives the credit Darwin receives. From their independent studies, the two eventually collaborated. The people in the other choices are associated with different accomplishments: Cuvier laid the foundations of comparative anatomy and vertebrate paleontology; Lamarck put forth the theory of evolution by the inheritance of aquired characteristics; Lyell provided evidence that the earth was millions to billions of years old, not, as was believed, about six thousand years old; Malthus was an economist whose ideas influenced the formation of Darwin's theory of natural selection.

63. (B)

If the frequency of A is .7, the frequency of a is .3, for the two must add up to 1.0, or 100%. Therefore, the frequency of aa is "a" squared. .3 squared is .09.

64. (B) 65. (C) 66. (A) 67. (D)

Answers to all of these come from basic knowledge of biologically important compounds. Carbohydrates are the main energy storers in organisms. Glucose (blood sugar), starch (energy storage in plants), and glycogen (energy storage in animals), are good examples. Lipids (fats) store about twice the amount of energy per gram as either carbohydrates or proteins do. Proteins, huge molecules built from amino acids, build body structures such as cell membranes, muscle tissue, etc. They are always a part of enzyme structure. DNA and RNA, the molecules involved in genetic expression, are nucleic acids. Choice E is not an appropriate match, as steroids are a separate family of molecules.

68. (D) 69. (C) 70. (A)

The most likely food chain whose members are of the group of listed organisms is:

alga ➡ minnow ➡ sunfish ➡ bass ➡ human

Photosynthetic algae build organic molecules and thus are the producers. A small-sized minnow is the most likely animal to feed on the plantlike algae as a primary consumer. As energy flows through a food chain, less useful energy remains available at each successive link. Humans are the last link in the chain and are the fourth-level consumers, therefore, in this food chain, the least amount of energy is available to the human.

71. (A) 72. (B) 73. (D) 74. (C)

Blood flows through blood vessels in the following sequence:

artery → arterioles → capillaries → venules → veins

Arteries are closest to the heart and its high blood pressure. Blood pressure slowly decreases as blood travels farther away from the heart. Veins, at the end of this route, return blood to the heart. Arterioles receive blood from arteries and supply it to the microscopic, numerous capillaries. At this level, cells are served. Nutrients and oxygen leave the blood through capillary walls to serve cells. Waste products and carbon dioxide leave body cells and enter blood for removal.

75. (D) 76. (E)

Chondrichthyes are the cartilagenous fish, e.g.: shark. Osteichthyes are the bony fish, e.g.: trout. The other choices belong to other vertebrate classes: bullfrog, Amphibians; rattlesnake, Reptiles; robin, Aves (birds).

77. (E) 78. (D) 79. (C) 80. (B)

Each is a definition for a particular example of symbiosis. A fox can hunt a rabbit (predation), whereas a tapeworm parasitizes a human body by living in the digestive tract. Bacteria receive a home while living in the human digestive tract and synthesize B vitamins for the human (mutualism). Other bacteria live on human skin, profiting from the relationship but having an apparently neutral effect on the human. Mutual suffering of organisms in a symbiotic relationship was not defined in any of the questions.

81. (A)

Structure A, the nucleolus, is the site of a great deal of RNA synthesis, especially the synthesis of rRNA (ribosomal RNA).

82. (E)

The endoplasmic reticulum (ER) offers a series of channels throughout the cell cytoplasm for transport. It thus serves as a microcirculatory system.

83. (C)

The centrioles are two cylinder-shaped organelles in the cytoplasm that are close to the nucleus. They are involved with spindle formation during cell reproduction. Its fibers attach to chromosomes, pulling and manipulating them during the division process. At all times, the centrioles are surrounded by short segments of microtubules, also referred to as astral rays.

84. (D)

This cell organelle is the mitochondrion. It is the site of cell respiration, where glucose is oxidized. Released energy is coupled to the formation of adenosine triphosphate, ATP. This universal energy currency is then spent by cells to run their metabolic activities: muscle contraction, active transport, etc.

85. (D)

Since the enzyme took the least amount of time (5 minutes) to convert one gram of substrate when the temperature was 37°C, its activity was greatest at this temperature. It can be seen, therefore that the enzyme's activity increased as the temperature increased, but peaked at a temperature of 37°C. It can also be seen that, at extremes of temperature (0°C and 70°C), enzyme activity is decreased.

Choice (B) is incorrect because it does not consider the enzyme's activity at 70°C in its statement.

86. (A)

Epidermis covers the outer surfaces of the leaf. Stomata are openings punctuating the epidermis for gas exchange. They are present throughout the lower covering but are absent in the upper layer of epidermis.

87. (B)

The parenchymal cells below the upper epidermal layer conduct photosynthesis. They are more densely packed, for higher photosynthetic efficiency, in the upper layer of palisade mesophyll. 100% humidity exists in layer C, the spongy mesophyll. Stomata are present in the underside of a leaf, and are labelled E in the diagram. The cuticle of a leaf is an outer layer that is secreted by cells in the layer that is labelled A, and serves to protect the plant from water loss.

88. (D)

All plant and animal proteins contain nitrogen (N). Nitrogen is part of the amino group present in all amino acids, which are the building blocks of all proteins. Plants can synthsize most proteins from simple inorganic ingredients that include nitrogen.

89. (D)

K is the symbol for potassium, which is the only nutrient among the five choices listed that is needed in large amounts and is therefore a macronutrient. Copper (Cu), magnesium (Mg), manganese (Mn), and zinc (Zn) are not needed in large amounts, and are therefore not macronutrients.

90. (B)

Respiratory function may be defined in terms of lung volumes and capacities. The respiratory rate is the number of breaths (inhalation and exhalation) per unit time. Tidal volume is the volume of air inspired or expired in an unforced respiratory cycle. Respiratory minute volume is the total volume of air exhaled per minute. Expiratory reserve volume is the maximum volume of air that can forcibly be exhaled immediately following tidal expiration. Inspiratory reserve volume is the maximum volume of air that can forcibly be inhaled immediately following tidal inspiration. Vital capacity is the maximum amount of air that can be expired following maximum inspiration. It is equal to the sum of the tidal volume, the expiratory reserve volume, and the inspiratory reserve volume. Thus, (A), the expiratory reserve volume, is equal to the difference of the vital capacity (5,000) and the inspiratory reserve volume and tidal volume (2500 + 500), which is 2000.

91. (B)

According to the definitions given above, vital capacity is equal to the sum of the tidal volume (400), the expiratory reserve volume (1500), and the inspiratory reserve volume (2400), which equal 4300.

92. (D)

The tidal volume multiplied by the respiratory rate is equal to the respiratory minute volume. Thus, the tidal volume is equal to the quotient of the respiratory minute volume (9000) divided by the rate (15), which is 600.

93. (C) 94. (A)

The portion of the graph that is labelled A depicts simple twitches. A simple twitch is a muscle contraction followed by total relaxation of the contracted muscle fibers.

Portion B represents the summation of many muscle twitches. At first, various muscle fibers contract. They then begin to relax. However, in summation, another twitch occurs before the muscle fibers have had enough time to totally relax. This produces more tension and force than would normally be produced by a lone muscle twitch. This increase in contraction is repeated, until the muscle fibers can contract no further. These muscle fibers are then in a state of tetanus, represented by portion C of the graph.

Fatigue, depicted by portion D, occurs because of a buildup of lactic acid in the muscle.

Tonus, the partial contraction of various muscle fibers during a period of relaxation, is not shown by the graph.

95. (B)

This leaf has one terminal point and net venation.

96. (A)

This leaf has more than one terminal point and palmate venation.

97. (E)

This leaf has more than one terminal point and palmately compound venation.

98. (D)

This leaf has one terminal point and parallel venation.

99. (D)

The period from noon to 2:00 p.m. consists of six twenty-minute intervals. Therefore, the size of the bacterial population will double six times by 2:00 p.m. $2^6 = 64$.

100. (C)

Bacterial cells would not lose their ability to reproduce in a fresh nutrient medium. If cell birth were greater, population size would increase (log phase). The opposite trend yields a population size decline (death phase). During the level stationary phase, the net growth of the size of the population is zero.

BIOLOGY

TEST III

BIOLOGY
ANSWER SHEET

1. (A) (B) (C) (D) (E)
2. (A) (B) (C) (D) (E)
3. (A) (B) (C) (D) (E)
4. (A) (B) (C) (D) (E)
5. (A) (B) (C) (D) (E)
6. (A) (B) (C) (D) (E)
7. (A) (B) (C) (D) (E)
8. (A) (B) (C) (D) (E)
9. (A) (B) (C) (D) (E)
10. (A) (B) (C) (D) (E)
11. (A) (B) (C) (D) (E)
12. (A) (B) (C) (D) (E)
13. (A) (B) (C) (D) (E)
14. (A) (B) (C) (D) (E)
15. (A) (B) (C) (D) (E)
16. (A) (B) (C) (D) (E)
17. (A) (B) (C) (D) (E)
18. (A) (B) (C) (D) (E)
19. (A) (B) (C) (D) (E)
20. (A) (B) (C) (D) (E)
21. (A) (B) (C) (D) (E)
22. (A) (B) (C) (D) (E)
23. (A) (B) (C) (D) (E)
24. (A) (B) (C) (D) (E)
25. (A) (B) (C) (D) (E)
26. (A) (B) (C) (D) (E)
27. (A) (B) (C) (D) (E)
28. (A) (B) (C) (D) (E)
29. (A) (B) (C) (D) (E)
30. (A) (B) (C) (D) (E)
31. (A) (B) (C) (D) (E)
32. (A) (B) (C) (D) (E)
33. (A) (B) (C) (D) (E)
34. (A) (B) (C) (D) (E)
35. (A) (B) (C) (D) (E)
36. (A) (B) (C) (D) (E)
37. (A) (B) (C) (D) (E)
38. (A) (B) (C) (D) (E)
39. (A) (B) (C) (D) (E)
40. (A) (B) (C) (D) (E)
41. (A) (B) (C) (D) (E)
42. (A) (B) (C) (D) (E)
43. (A) (B) (C) (D) (E)
44. (A) (B) (C) (D) (E)
45. (A) (B) (C) (D) (E)
46. (A) (B) (C) (D) (E)
47. (A) (B) (C) (D) (E)
48. (A) (B) (C) (D) (E)
49. (A) (B) (C) (D) (E)
50. (A) (B) (C) (D) (E)
51. (A) (B) (C) (D) (E)
52. (A) (B) (C) (D) (E)
53. (A) (B) (C) (D) (E)
54. (A) (B) (C) (D) (E)
55. (A) (B) (C) (D) (E)
56. (A) (B) (C) (D) (E)
57. (A) (B) (C) (D) (E)
58. (A) (B) (C) (D) (E)
59. (A) (B) (C) (D) (E)
60. (A) (B) (C) (D) (E)
61. (A) (B) (C) (D) (E)
62. (A) (B) (C) (D) (E)
63. (A) (B) (C) (D) (E)
64. (A) (B) (C) (D) (E)
65. (A) (B) (C) (D) (E)
66. (A) (B) (C) (D) (E)
67. (A) (B) (C) (D) (E)
68. (A) (B) (C) (D) (E)
69. (A) (B) (C) (D) (E)
70. (A) (B) (C) (D) (E)
71. (A) (B) (C) (D) (E)
72. (A) (B) (C) (D) (E)
73. (A) (B) (C) (D) (E)
74. (A) (B) (C) (D) (E)
75. (A) (B) (C) (D) (E)
76. (A) (B) (C) (D) (E)
77. (A) (B) (C) (D) (E)
78. (A) (B) (C) (D) (E)
79. (A) (B) (C) (D) (E)
80. (A) (B) (C) (D) (E)
81. (A) (B) (C) (D) (E)
82. (A) (B) (C) (D) (E)
83. (A) (B) (C) (D) (E)
84. (A) (B) (C) (D) (E)
85. (A) (B) (C) (D) (E)
86. (A) (B) (C) (D) (E)
87. (A) (B) (C) (D) (E)
88. (A) (B) (C) (D) (E)
89. (A) (B) (C) (D) (E)
90. (A) (B) (C) (D) (E)
91. (A) (B) (C) (D) (E)
92. (A) (B) (C) (D) (E)
93. (A) (B) (C) (D) (E)
94. (A) (B) (C) (D) (E)
95. (A) (B) (C) (D) (E)
96. (A) (B) (C) (D) (E)
97. (A) (B) (C) (D) (E)
98. (A) (B) (C) (D) (E)
99. (A) (B) (C) (D) (E)
100. (A) (B) (C) (D) (E)

BIOLOGY EXAM III

PART A

Directions: For each of the following questions or incomplete sentences, there are five choices. Choose the answer which is most correct. Darken the corresponding space on your answer sheet.

1. All of the following are recycled on earth EXCEPT:

   (A) carbon
   (B) oxygen
   (C) energy
   (D) water
   (E) nitrogen

2. Agonistic behavior may be expressed as all of the following EXCEPT:

   (A) threat
   (B) appeasement
   (C) displacement
   (D) kinesis
   (E) attack

3. A group of potentially interbreeding organisms that may or may not live in the same area is called a

   (A) population
   (B) species
   (C) community
   (D) ecosystem
   (E) race

4. The table below represents the composition of bases present in DNA extracted from different organisms.

	base composition (percent)			
Source	Adenine	Guanine	Cytosine	Thymine
human	31.0	19.6	19.3	30.1
mouse	29.0	21.1	21.2	28.7
tobacco	29.7	20.0	20.4	29.9
E. coli	24.7	25.3	25.4	24.6
sea urchin	32.7	17.7	17.3	32.3

From the table above, which of the following can be concluded?

(A) The nucleotide bases are bound by hydrogen bonds.

(B) The bases are attached to a sugar-phosphate backbone.

(C) The base composition of DNA is the same in all organisms.

(D) The amount of adenine equals the amount of thymine.

(E) DNA is a double helix.

5. Which of the following does NOT match?

(A) roots - nutrient storage

(B) xylem - movement of water

(C) phloem - movement of sap

(D) leaves - nitrogen uptake

(E) stem tip - growth

6. Phosphorus is made available to living organisms because of

   (A) photosynthesis

   (B) legumes

   (C) upwellings of water from the bottoms of lakes and oceans

   (D) cellular respiration

   (E) water runoff

7. Amino acids that must be taken in by the body in food are all of the following EXCEPT:

   (A) also made by the body

   (B) necessary for growth

   (C) found in the proteins that are eaten

   (D) basic components of our body proteins

   (E) as important as nonessential amino acids

8. The ability of a plant to survive on land is not increased when that land plant

   (A) produces free-swimming sperm

   (B) has a cuticle on the parts exposed to the air

   (C) contains xylem

   (D) makes seeds

   (E) has roots

9. When a young bird learns to follow the first large, moving object it sees and hears, this is an example of

(A) trial-and-error learning
(B) imprinting
(C) habituation
(D) taxis
(E) kinesis

10. All of the following statements apply to carbohydrates EXCEPT:

(A) they contain C:H:O in a 1:2:1 ratio
(B) they are the primary energy source for living things
(C) they are made of glycerol
(D) they may form polysaccharides
(E) they are organic compounds

11. Territories are areas that increase opportunities for food procurement, mating, and/or nesting sites. They are defended by the occupants against others, usually of the same species. Territories serve to

(A) provide for equal distribution of food
(B) increase the time that individuals spend fighting
(C) regulate population size
(D) increase physical contact among species members
(E) decrease reproductive success

12. Heterotrophic bacteria and fungi are similar in that both

(A) possess food vacuoles

(B) have gastrovascular cavities

(C) undergo extracellular digestion

(D) have cilia for filter feeding

(E) undergo intracellular digestion

13. The children of a father with type O blood and a mother with type AB blood could have the blood type

(A) O

(B) AB

(C) O or AB

(D) A or B

(E) O, A, B, or AB

14. The ecological role played by trees on land is the same as that played by which organisms in the ocean?

(A) bacteria

(B) algae

(C) fish

(D) shrimp

(E) sharks

15. All of the following are true about genes EXCEPT:

(A) they are found on chromosomes

(B) they contain DNA

(C) they exist in different forms

(D) they can undergo change

(E) they are always beneficial

16. Because whales are classified as mammals, you would expect them to have all of the following characteristics EXCEPT:

(A) a vertebral column

(B) a dorsal hollow nerve cord

(C) scales

(D) warm-bloodedness

(E) ability to nurse their young

17. A cross is made between a wild-type female fruit fly and a miniature-winged male fly. All of the offspring in the first generation are wild-type. In the second generation, 1/4 of the offspring is miniature-winged and male. From this we can conclude that the gene for miniature-winged flies is

(A) dominant and autosomal

(B) recessive and autosomal

(C) dominant and X-linked

(D) recessive and X-linked

(E) recessive and Y-linked

18.

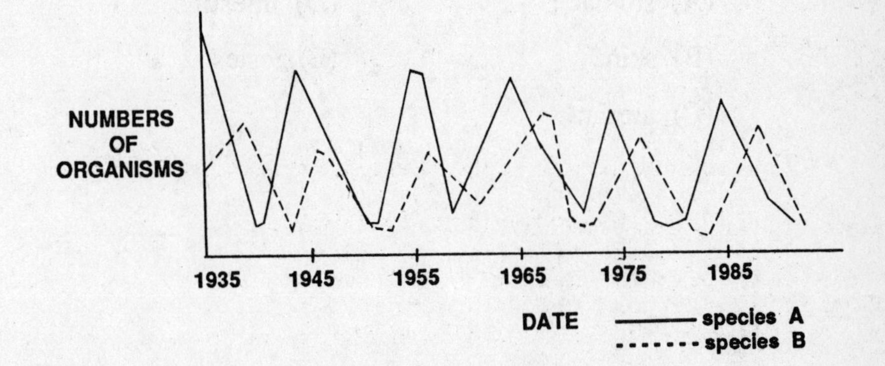

Species A and species B are members of the same community. From this graph we can conclude that

(A) species A eats species B

(B) species A increases in number every 5 years

(C) species B eats species A

(D) species B can increase in number when A is low

(E) both species are competing for the same food source

19. Which of the following is NOT an energy-requiring process?

(A) operation of the sodium-potassium pump

(B) production of glycogen from glucose

(C) muscle contraction

(D) movement of water across a membrane

(E) uptake of calcium against a concentration gradient

20. In humans, cells with the haploid number of chromosomes occur in the

(A) stomach

(B) skin

(C) urethra

(D) intestine

(E) testes

21. An enzyme is all of the following EXCEPT:

(A) a catalyst

(B) made by living organisms

(C) used up during a chemical reaction

(D) specific in its effect

(E) a protein

22. Two organisms that occupy the same niche would NOT be expected to

(A) mate at different times of the year

(B) compete for the same food

(C) behave in a similar manner

(D) live in the same place

(E) survive best in a similar environment

23. The formation of a peptide bond involves

I. two monosaccharides

II. two amino acids

III. a condensation reaction

(A) I only

(B) II only

(C) III only

(D) I and III

(E) II and III

24. Blood pressure decreases as the blood moves farther away from the heart. This drop in pressure is due to the fact that

(A) the blood moves against gravity in the veins

(B) the heart must first pump to the lungs

(C) valves are present in the arteries

(D) the blood is slowed down by friction between it and the blood vessel walls

(E) arteries are larger in diameter than veins

25. Implantation of the embryo normally occurs in the

(A) ovary

(B) fallopian tube

(C) uterus

(D) vagina

(E) abdominal cavity

26. A bacterial gene 450 bases long will produce a protein containing approximately how many amino acids?

(A) 150

(B) 300

(C) 450

(D) 900

(E) 1350

27. Which do reptiles and amphibians have in common?

(A) Egg-laying in moist or wet environments

(B) A four-chambered heart

(C) Lungs in the adult form

(D) External fertilization

(E) Dry, scaly skin

28. Natural selection can occur because

(A) fossils have been found

(B) the limbs of amphibians, reptiles, birds and mammals are similar in structure

(C) all living things contain DNA

(D) more organisms are produced than can survive

(E) extinction decreases genetic variability

29. Which of the following plant modifications does NOT contribute to water retention?

(A) closing of stomata by guard cells in the leaves

(B) movement of water in xylem

(C) waxy cuticle on surfaces of the plant parts

(D) roots

(E) leaf shape

30. Which of the following does NOT match?

(A) transfer RNA - amino acid transport

(B) nucleus - site of RNA production

(C) ribosomes - translation

(D) gene - protein assembly

(E) messenger RNA - copy of gene

31. The ecosystem characterized by coniferous trees is the

(A) temperate forest
(B) tundra
(C) taiga
(D) chaparral
(E) tropical rain forest

32. Evolutionary change may occur in all of the following situations EXCEPT:

(A) some bacteria survive after exposure to antibiotics because they are antibiotic-resistant
(B) a small number of individuals survive a hurricane that kills most of the people on an island
(C) a species of birds with a variety of genetically different forms migrates to a new island
(D) dogs are selected for breeding because of certain behavioral characteristics that they did not acquire during training
(E) horses raised on a nutritionally well-balanced diet are selected for breeding

33. In the grasslands, trees do not replace the grasses as part of an ecological succession because of

(A) insects
(B) water limits and fire
(C) limited sunlight
(D) cool temperatures
(E) the type of soil

34. Simple proteins and DNA are similar in that they both contain all of the following EXCEPT:

(A) oxygen
(B) hydrogen
(C) nitrogen
(D) phosphorus
(E) carbon

35. Bacteria contain

(A) true nuclei
(B) chloroplasts
(C) mitochondria
(D) simple, circular DNA
(E) cell membranes as their outermost structures

36. The autonomic nervous system is characterized by involuntary control of bodily functions by motor neurons. All of the following are examples of autonomic function EXCEPT:

(A) stimulation of muscles of the digestive tract
(B) increased secretion of salivary glands
(C) contraction of skeletal muscles
(D) release of adrenalin by the adrenal medulla
(E) increased heart rate in response to an emergency

37. One important reason for classifying fungi in a separate kingdom from plants is that fungi

(A) may be multicellular

(B) are eukaryotic

(C) are nonphotosynthetic

(D) produce spores

(E) have cell walls

38. Which of the following does NOT have the ability to reproduce asexually?

(A) yeast

(B) hydra

(C) bacterium

(D) bird

(E) *Paramecium*

39. In pea plants, the round seed allele is dominant over the wrinkled seed allele, and the yellow seed allele is dominant over the green seed allele. The genes for seed texture and those for seed color are on different chromosomes. A plant heterozygous for seed texture and seed color is crossed with a plant that is wrinkled and heterozygous for seed color. What phenotypic ratio of offspring would you expect in the next generation?

(A) 3:1

(B) 3:3:1:1

(C) 9:3:3:1

(D) 1:1

(E) 1:1:1:1

40. The greatest number of mitotic divisions occurs in

(A) meristematic tissue
(B) xylem
(C) phloem
(D) guard cells
(E) cork

41. Where there is increased parental involvement in the raising of offspring, we would expect

(A) large numbers of offspring
(B) a shorter time before maturation
(C) little learning by the offspring
(D) eating of the young
(E) the welfare of the young placed before the welfare of the parent

42. A single base change in the nucleotide sequence of a gene will

(A) always result in a change in the protein produced
(B) be harmful
(C) change the length of the gene
(D) affect a critical portion of the protein
(E) be inherited

43. Protozoa

(A) are multicellular

(B) are prokaryotic

(C) can make their own food

(D) have a variety of techniques for locomotion

(E) do not have membrane-bound internal structures

44. Which of the following are normally diploid?

(A) spores

(B) eggs

(C) drones (male bees)

(D) zygotes

(E) sperm

45. As one moves up an ecological pyramid, generally

(A) the biomass increases

(B) photosynthesis increases

(C) energy levels decrease

(D) the number of organisms increases

(E) productivity increases

46. Cellular respiration is a series of chemical reactions that break down organic materials in the body to extract energy. When the body temperature of an organism increases, we would expect

(A) the rate of cellular respiration to increase

(B) less energy production

(C) the formation of more enzymes

(D) no change in cellular respiration or energy production

(E) a decrease in the ability to perform other energy dependent activities.

47. If two genes are linked

(A) they will always be inherited together

(B) they will be inherited together unless crossing over has occurred

(C) the frequency of chromosomal rearrangements will increase

(D) independent assortment will occur

(E) blending inheritance will occur

48.

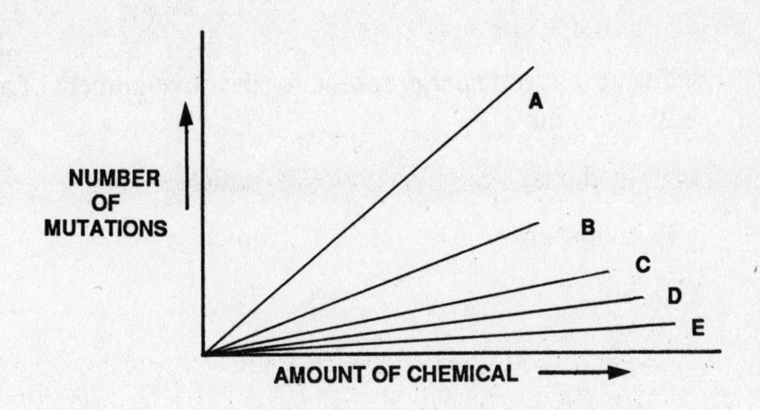

From the graph above, we can conclude that

(A) chemical A causes the most mutations

(B) all chemicals cause the same number of mutations

(C) as the amount of chemical increases the number of mutations decreases

(D) high concentrations of a chemical always result in large numbers of mutations

(E) mutations are harmful

49. How many different gametes will be produced by an organism that has the genotype AaBbCCDd for four genes that assort independently?

(A) 2

(B) 3

(C) 4

(D) 6

(E) 8

50. The nutrient and energy source for the development of a plant embryo is the

(A) seed coat

(B) endosperm

(C) fruit

(D) radicle

(E) epicotyl

51. Which of the following depends upon a high degree of communications?

I. social behavior
II. echolocation
III. circadian rhythms

(A) I only
(B) II only
(C) III only
(D) I and II
(E) I and III

52. Genetic variability can result from all of the following EXCEPT:

(A) mutation
(B) recombination
(C) migration
(D) natural selection
(E) chromosomal rearrangement

53. When species A, which is not protected by an unpleasant characteristic, looks like (mimics) species B, which has an unpleasant taste, we can predict that

(A) species A will be readily preyed upon
(B) species A will not be readily preyed upon
(C) species B will be readily preyed upon
(D) species A and species B will be readily preyed upon
(E) neither species will affect the survival of the other

54. All of the following are involved in movement EXCEPT:

(A) cilia
(B) flagella
(C) actin and myosin
(D) spindle fibers
(E) lysosomes

55. Foxes may eat rabbits and, therefore, foxes are

(A) parasites
(B) predators
(C) decomposers
(D) herbivores
(E) producers

56. When the heterozygotes in a population have a reproductive advantage, we can predict that

(A) they will produce only heterozygous offspring
(B) the individuals that are homozygous dominant will take over
(C) the individuals that are homozygous recessive will take over
(D) both homozygous dominant and homozygous recessive individuals will continue to appear in future generations
(E) homozygous dominant and homozygous recessive individuals will be eliminated

57. Angiosperms are unique in that they

(A) are seed-producing plants
(B) have vascular tissue
(C) have a gametophyte that is retained in the sporophyte
(D) have true roots and leaves
(E) produce flowers and fruit

58.

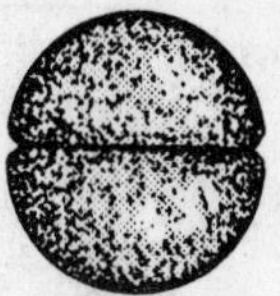

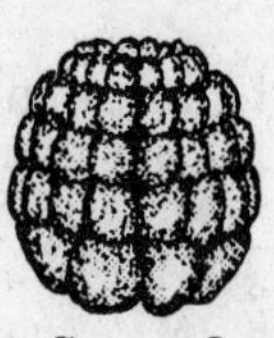
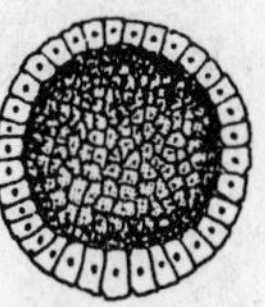
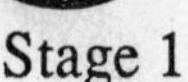

Stage 1 Stage 2 Stage 3 Stage 4

The next stage in the process represented in the drawings above is called:

(A) early cleavage

(B) neural tube formation

(C) gastrulation

(D) blastula formation

(E) morula formation

PART B

DIRECTIONS: Each of the following sets of questions refers to a list of lettered choices immediately preceding the question set. Choose the best answer from the list, then darken the corresponding space on your answer sheet. Choices may be used once, more than once, or not at all.

Questions 59 - 62

(A) mitochondrion

(B) endoplasmic reticulum

(C) cytoplasm

(D) chloroplasts

(E) vacuoles

59. Site of glycolysis

60. Serve(s) a variety of functions, including water and glucose storage in plants

61. Place where NADH transfers its high-energy electrons to make ATP

62. Covert(s) light energy into chemical energy

Questions 63 - 65

(A) anther

(B) stigma

(C) ovary

(D) petal

(E) sepal

63. The site of pollen formation

64. The portion of the flower that attracts insects that aid in pollination

65. The place in which the sperm nucleus fuses with the egg nucleus

Questions 66 - 68

(A) Cnidarians

(B) Echinoderms

(C) Platyhelminthes

(D) Annelids

(E) Arthropods

66. Have radial symmetry, but no coelom

67. Are segmented worms

68. Have a jointed exoskeleton

Questions 69 - 70

(A) ligament
(B) cartilage
(C) nerves
(D) blood vessels
(E) tendon

69. Connects bone to bone

70. Provides bone with oxygen and nutrients

Questions 71 - 74

(A) cellulose
(B) carbon dioxide
(C) ATP
(D) oxygen
(E) glucose

71. A carbon compound that is made during photosynthesis

72. A gaseous end-product of photosynthesis

73. A cell wall component

74. The energy currency of most living cells, including plants

Questions 75 - 78

(A) salivary glands
(B) gall bladder
(C) stomach
(D) liver
(E) pancreas

75. Begins starch digestion

76. Produces pepsin

77. The site of bile production

78. Produces insulin as well as some digestive enzymes

PART C

Directions: The following questions refer to the accompanying diagram. Certain parts of the diagram have been labelled with numbers. Each question has five choices. Choose the best answer and darken the corresponding space on your answer sheet.

Questions 79 - 83

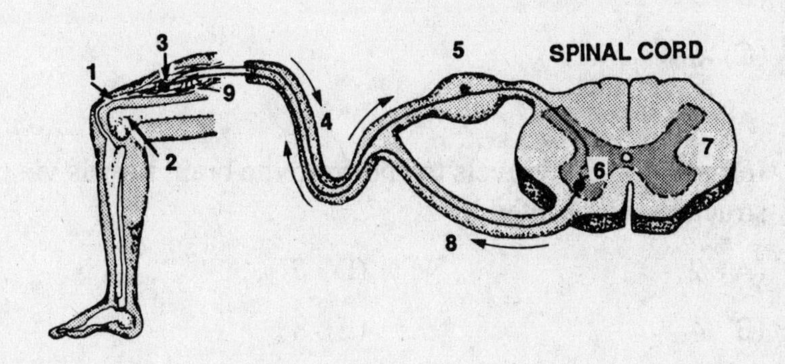

79. The sensory receptor that picks up information about the muscle lengthening is represented by

(A) 1
(B) 2
(C) 3
(D) 5
(E) 6

80. The nucleus of the sensory neuron is located at the site represented by

(A) 3
(B) 5
(C) 6
(D) 7
(E) 9

81. A synapse is represented by

(A) 4
(B) 5
(C) 6
(D) 7
(E) 8

82. Information that travels to and from the brain travels via the structure represented by

(A) 2
(B) 4
(C) 5
(D) 7
(E) 8

83. The place at which the muscle is stimulated to contract by the motor neuron is designated by

(A) 2
(B) 3
(C) 5
(D) 6
(E) 9

PART D

DIRECTIONS: Each of the following questions refer to a laboratory or experimental situation. Read the description of each situation, then select the best answer to each question. Darken the corresponding space on the answer sheet.

Questions 84 - 86 refer to the four graphs below. Graphs I, II, III, and IV show the growth of plant cells in the presence of different concentrations of the plant hormone kinetin. In addition, Graphs II, III, and IV show the growth of plant cells in the presence of three different concentrations of the plant hormone auxin.

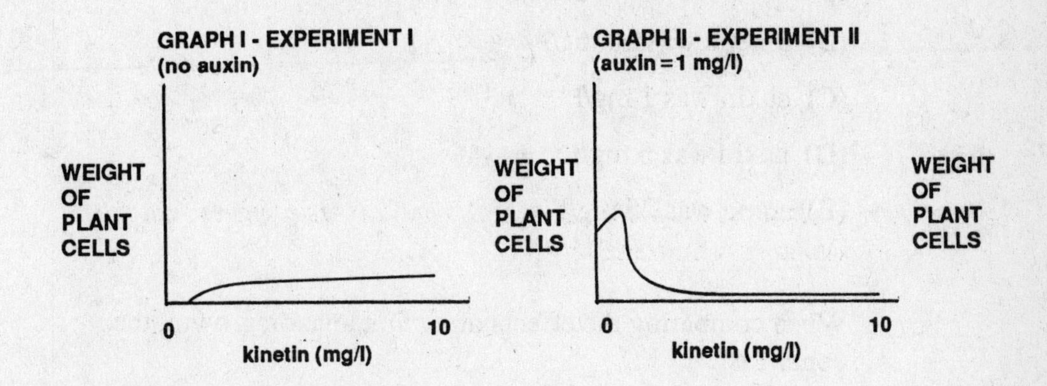
GRAPH I - EXPERIMENT I
(no auxin)
WEIGHT
OF
PLANT
CELLS
0
10
kinetin (mg/l)
GRAPH II - EXPERIMENT II
(auxin =1 mg/l)
WEIGHT
OF
PLANT
CELLS
0
10
kinetin (mg/l)
WEIGHT
OF
PLANT
CELLS

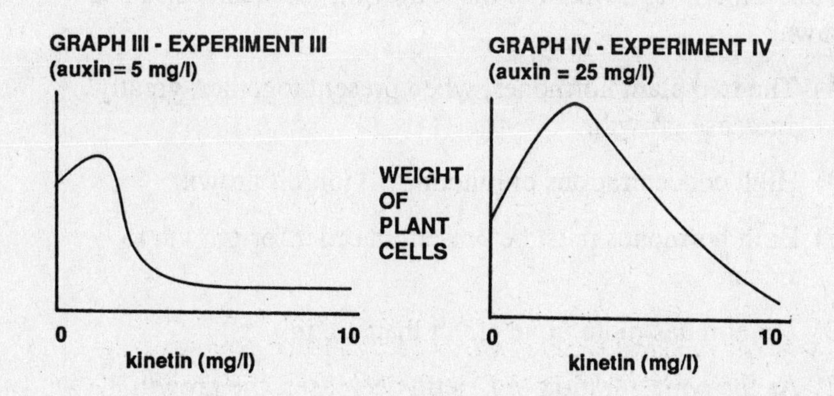
GRAPH III - EXPERIMENT III
(auxin= 5 mg/l)
0
10
kinetin (mg/l)
GRAPH IV - EXPERIMENT IV
(auxin = 25 mg/l)
WEIGHT
OF
PLANT
CELLS
0
10
kinetin (mg/l)

84. The optimum growth occurred when the concentration of

(A) kinetin was 0 mg/l

(B) kinetin was 10 mg/l

(C) auxin was 1 mg/l

(D) auxin was 5 mg/l

(E) auxin was 25 mg/l

85. When comparing the effects of auxin alone on growth, the control is

(A) not present

(B) graph I

(C) graph II

(D) graph III

(E) graph IV

86. In this experiment, which of the following conclusions can be drawn?

(A) The two plant hormones, when present together, greatly increase growth.

(B) High concentrations of kinetin will inhibit growth.

(C) Both hormones must be present in order for growth to occur.

(D) Kinetin has more of an effect than auxin.

(E) As the concentration of kinetin decreases, the growth increases.

Questions 87 - 88 refer to the diagram of the cell membrane below.

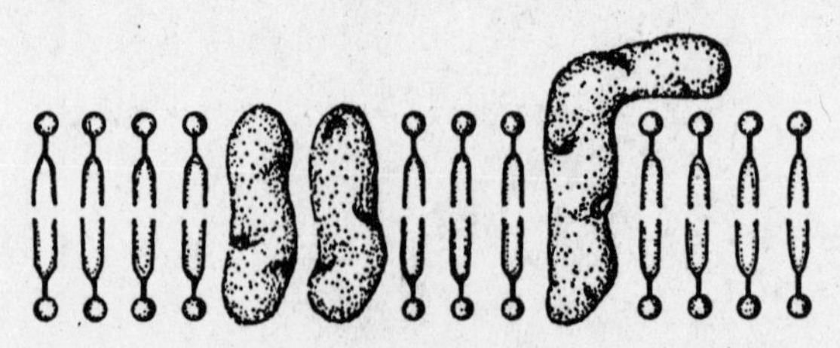

87. The primary component of this cell membrane is

(A) carbohydrate
(B) phospholipid
(C) nucleic acid
(D) protein
(E) salt

88. Which of the following would NOT be considered true about this membrane?

(A) It is most permeable to proteins.
(B) Salts pass through pores in the membrane.
(C) There is a positive charge on the outside.
(D) Hormones and other chemicals may bind with outside receptors.
(E) Movement of materials across the membrane may require energy.

Questions 89 - 92

component	lymph	blood	urine	cerebro-spinal fluid	intercellular fluid
protein	4000	7000	0	20000	0.02
glucose	100	100	0	64	100
urea	15	15	26	12	15
sodium	14.1	14.1	5-13	14.7	14.2
potassium	0.47	0.5	2-7	0.2	0.48

(amount present - relative units)

89. The fluid with the greatest variability in amounts of materials present is

(A) lymph
(B) blood
(C) urine
(D) cerebrospinal fluid
(E) intercellular fluid

90. From this data we can conclude that

(A) lymph, blood and intercellular fluid are similar except for protein content
(B) glucose is normally excreted by the body
(C) cerebrospinal fluid is the same as blood
(D) the largest component in urine is protein
(E) the largest component in lymph is glucose

91. In addition to the components listed, lymph also contains

(A) red blood cells
(B) white blood cells
(C) albumin
(D) platelets
(E) fibrinogen

92. Some nitrogenous wastes are removed from the body as

(A) protein
(B) urea
(C) glucose
(D) sodium
(E) potassium

Questions 93 and 94

Briggs and King and other researchers used a technique for transferring nuclei from embryonic cells into amphibian eggs whose nuclei had been removed. Nuclei were taken from blastula, gastrula and intestinal cells and injected into the eggs. Cytoplasm by itself was also injected into some eggs. Many normal adult frogs developed from the eggs containing blastula nuclei. A few normal adult frogs developed from the eggs containing gastrula nuclei or intestinal nuclei. Eggs injected with cytoplasm alone did not develop normally.

93. Which of the above was the control?

(A) cytoplasm alone

(B) blastula nucleus

(C) gastrula nucleus

(D) intestinal cell nucleus

(E) there was no control

94. This experiment shows that

(A) the nucleus supports development

(B) the egg rejects the nucleus during an immune reaction

(C) genes are lost during development

(D) mutations occur in the genes in the nucleus

(E) the nucleus remains the same throughout development

Questions 95 - 96

Samples of blood were taken from three different individuals and chemically analyzed for different forms of hemoglobin. Individuals A and C are phenotypically normal and individual B suffers from sickle cell anemia. The chart below shows the types of hemoglobin for the three different individuals.

	forms of hemoglobin	
individual	type 1	type 2
A	+	-
B	-	+
C	+	+

(+ present, - absent)

95. Which of the following statements about the experiment above is correct?

(A) A is heterozygous.

(B) B is heterozygous.

(C) C is heterozygous.

(D) A person with sickle cell anemia carries a normal gene for hemoglobin.

(E) The gene for sickle cell anemia behaves as a dominant gene when we look at the expression of disease.

96. The gene for sickle cell anemia is different from the normal hemoglobin gene, because

(A) it contains ribose

(B) it is missing

(C) it has a base substitution

(D) it is made of RNA instead of DNA

(E) it contains uracil

Questions 97 - 100 refers to the growth of algae in a lake. The graph below shows the number of algae present throughout the year, as well as changes in nutrients, light, and temperature.

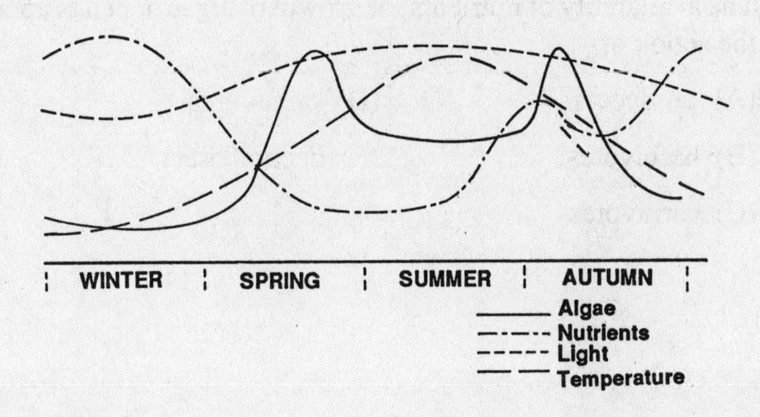

97. The rate of growth of algae is greater in the

(A) winter

(B) spring and summer

(C) spring and autumn

(D) winter and summer

(E) summer and autumn

98. During the summer, growth of algae is limited by

(A) light

(B) temperature

(C) light and temperature

(D) nutrients

(E) light, temperature, and nutrients

Peaks in growth of algae depend upon

(A) light alone

(B) temperature alone

(C) nutrients alone

(D) light and temperature together

(E) light, temperature, and nutrients

100. The availability of nutrients for growth of algae depends upon the action of

(A) producers

(B) herbivores

(C) carnivores

(D) omnivores

(E) decomposers

BIOLOGY
TEST III

ANSWER KEY

1.	C	34.	D	67.	D
2.	D	35.	D	68.	E
3.	B	36.	C	69.	A
4.	D	37.	C	70.	D
5.	D	38.	D	71.	E
6.	C	39.	B	72.	D
7.	A	40.	A.	73.	A
8.	A	41.	E	74.	C
9.	B	42.	E	75.	A
10.	C	43.	D	76.	C
11.	C	44.	D	77.	D
12.	C	45.	C	78.	E
13.	D	46.	A	79.	C
14.	B	47.	B	80.	B
15.	E	48.	A	81.	C
16.	C	49.	E	82.	D
17.	D	50.	B	83.	E
18.	C	51.	A	84.	E
19.	D	52.	D	85.	B
20.	E	53.	B	86.	A
21.	C	54.	E	87.	B
22.	A	55.	B	88.	A
23.	E	56.	D	89.	C
24.	D	57.	E	90.	A
25.	C	58.	C	91.	B
26.	A	59.	C	92.	B
27.	C	60.	E	93.	A
28.	D	61.	A	94.	A
29.	B	62.	D	95.	C
30.	D	63.	A	96.	C
31.	C	64.	D	97.	C
32.	E	65.	C	98.	D
33.	B	66.	A	99.	E
				100.	E

BIOLOGY
TEST III

DETAILED EXPLANATIONS OF ANSWERS

1. (C)

This question deals with biogeochemical cycles in ecology. Matter can be transferred from living material to the physical environment and back. Thus, carbon, oxygen, water, and nitrogen (matter) can be recycled. The only way that living material can obtain energy is by photosynthesis. Light energy from the sun is converted to chemical energy. As this chemical energy is transferred along the food chain, chemical bonds are broken and reformed, and heat energy is produced. Much of this heat energy is lost to the environment and cannot be recovered by living organisms.

2. (D)

In order to answer this question you must know what is meant by agonistic behavior and the forms it may take. Agonistic behavior is aggressive behavior. It may manifest itself in its most serious form as an attack on another organism. It may also be expressed as displays. Threat displays and appeasement displays are ways that animals have of resolving a conflict without actual combat. The animal with the strongest threat displays (larger appearance, menacing posture) usually wins. The loser signals the end of the conflict with appeasement behavior (minimizing size, showing the most vulnerable part of the body, turning away). Displacement behavior is an irrelevant response to a situation. This may occur in response to aggressive behavior, but may also occur in other situations. A kinesis is a stable pattern of behavior that is an undirected response to a simple stimulus, and, thus, has nothing to do with agonistic behavior.

3. (B)

This is a straightforward question that tests your knowledge of ecological concepts. An ecosystem includes all the organisms that live in a particular area and their physical environment. Those organisms would make up a community.

Population, species, and race are terms that define a group of potentially interbreeding organisms. However, population and race refer to organisms inhabiting the same area. In fact, a race (also called a subspecies) is a type of population that is geographically more isolated and, therefore, has become genetically distinct.

A species is a group of potentially interbreeding organisms, whether or not they live in the same area. Thus, the term species is more inclusive than the terms populations and races.

4. (D)

For this question you must be able to interpret the data given in tabular form. While a number of true statements are offered as choices (nucleotide bases are hydrogen bonded, bases are attached to a sugar-phosphate backbone, and DNA is a double helix), none of these can be concluded from the information given in the question. Looking at the percent compositions for each organism, we can see that the base composition may be different. When we compare percentages of bases, we see that the number of adenines equals the number of thymines.

5. (D)

This question deals with the structure and function of plant parts. Roots, among other functions, are responsible for the storage of nutrients made by the plant. Movement of water occurs via xylem, while movement of sap occurs via phloem. Meristematic tissue is where cell division occurs, and one location for this tissue is the tip of the stem.

While the nitrogen content of air is about 78%, gaseous nitrogen cannot be used by plants. Therefore, nitrogen uptake does not occur through the leaves. Nitrate and ammonia ions are formed by microorganisms in the soil from gaseous nitrogen during the process of nitrogen fixation. Nitrate and ammonium ions may also be supplied by decomposers. These ions are then taken up by the plant roots.

6. (C)

This question concerns nutrient cycles, and, in particular, the phosphorus cycle. The carbon cycle includes photosynthesis and cellular respiration. Photosynthesis binds carbon from carbon dioxide into the carbon compound glucose, while cellular respiration, the process of energy production, breaks down glucose forming carbon dioxide. Legumes are plants that contain bacteria in their roots and participate in the nitrogen cycle. The bacteria undergo nitrogen fixation, converting gaseous nitrogen into nitrogen compounds that can be used by the plants. The water cycle includes the movement of water from land to lakes, rivers, and oceans as water runoff.

Phosphorus, which is lost to living organisms when it is washed away with water runoff, can also be recycled. Water at the bottom of a lake or ocean can upwell. When the temperature of the bottom water is higher than surface water temperatures, the bottom water rises, taking settled nutrients, including phosphorus, with it. Phosphorus is then available for incorporation into plants growing near the surface of the particular water body.

7. (A)

In order to answer this question you must be familiar with the formation of proteins and the role of proteins in the body. Twenty different amino acids are needed by vertebrates to make proteins, important compounds necessary for growth. Humans make twelve of these amino acids, the so-called nonessential ones. However, the eight essential amino acids can only be supplied through the diet. Essential and nonessential amino acids must be present together in order for protein synthesis to occur.

8. (A)

To answer this question correctly you must understand that land plants must retain water in order to survive. They have, therefore, developed mechanisms to minimize water loss. All land plants have a cuticle on all parts exposed to the air to prevent water loss by evaporation. In addition, the presence of roots for water uptake, as well as vascular tissue (such as xylem) for the distribution of water and other materials, increases the ability of the plant to survive on land. Seed production among gymnosperms and angiosperms further liberates plants from a water environment. Seeds with their protective seed coats provide plants with a nonaqueous means of reproduction. With the development of seeds, free-swimming sperm have been replaced by pollen grains which contain a generative nucleus. The generative nucleus divides to produce sperm. Thus, pollen grains eliminate the need for external water for fertilization.

9. (B)

For this question you must be familiar with animal behavior, and, in particular, different types of learning. A taxis is an oriented movement in response to an environmental stimulus, while a kinesis is a random response. Neither of these are learned behaviors.

Trial-and-error, habituation, and imprinting are all considered types of learning. Trial-and-error learning involves improvement in a response based upon previous experience. Habituation, on the other hand, is a decreased response to a stimulus after continued exposure to the stimulus. The behavior of the young bird is an example of imprinting. This type of learning occurs during a sensitive period in the development of the organism and usually is irreversible.

10. (C)

For this question you must know the structural and functional characteristics of carbohydrates. Carbohydrates are organic compounds, or carbon compounds. They contain carbon, hydrogen, and oxygen in a 1:2:1 ratio. The simple sugars are the basic unit of carbohydrates, usually containing three to six carbon atoms. These simple sugars, or monosaccharides, may be combined to form starches, or polysaccharides. Carbohydrates, once broken down, are energy sources for living things. Glycerol is not a raw material for carbohydrates, but rather is necessary for the formation of fats.

11. (C)

This question tests your ability to understand what a territory is, as defined in the question, and to apply that knowledge to understand its ecological functions. A territory is a particular area that is defended by its occupant, usually from others of the same species. Since the individual defends the borders of the territory, the territory actually decreases physical contact among species members. It therefore, decreases the time and energy devoted to aggression.

A territory channels resources available to specific individuals, maximizing their chances of surviving and reproducing. It may insure a food supply for the individual, while limiting food available to those who have not been successful in establishing one. It also increases the reproductive success of the individual in it, by increasing the chance of mating. Therefore, it increases the reproductive success of the entire species. Equal distribution of resources among too many individuals could compromise the ability of any of them to survive. Therefore, the territory is a way of regulating population size.

12. (C)

This question asks you to look at a variety of ways that organisms obtain nutrients, and determine what bacteria and fungi have in common. Food vacuoles form by phagocytosis, and are a means of engulfing food material for later digestion within the vacuole. This type of food procurement is characteristic of amoebae. Food may also be digested in a cavity called the gastrovascular cavity. Planaria and coelenterates have gastrovascular cavities. Clams trap food in mucus that flows along its gills. The trapped food is moved by cilia to the mouth for entry into the internal digestive system, a process called filter feeding. Food vacuole formation is a form of intracellular digestion, while the gastrovascular cavity and filter feeding provide for extracellular digestion.

Fungi and bacteria do not have internal digestive systems. They, therefore, rely upon excretion of digestive enzymes onto living or dead organic material, and absorption of the digestive end-products. Thus, digestion in fungi and bacteria is extracellular.

13. (D)

In order to predict the outcome of a mating between a type O father and a type AB mother, you must know something about the forms of the gene that determine the ABO blood types. Gene I^A is a dominant gene that codes for type A antigens, proteins found on red blood cells. Gene I^B is also a dominant gene and codes for type B antigens found on red blood cells. Gene i is the recessive form that gives neither type A antigens nor type B antigens. A father with type O blood has the genotype ii, and can only produce i sperm. A mother with type AB blood has the genotype I^AI^B, and can produce either I^A or I^B eggs. An i sperm will fertilize the I^A egg to produce a child with an I^Ai genotype, or an i sperm will fertilize the I^B egg to produce a child with an I^Bi genotype. Thus, the children of that mating can have either type A or type B blood.

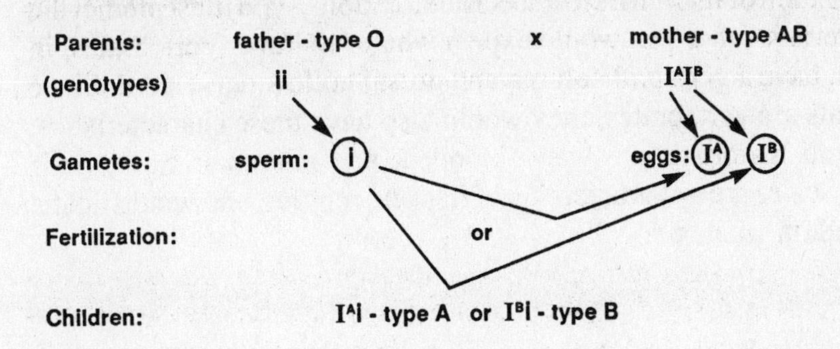

14. (B)

For this question you must know what role trees play in the ecosystem and determine what organism plays the same role in the ocean. Trees are primary producers. They have the ability to make organic molecules from inorganic molecules using sunlight as the energy source. In other words, they are photosynthetic organisms. Algae also undergo photosynthesis and, therefore, play the same role as trees in their ecological community. Fish, shrimp, sharks, and bacteria are consumers and/or decomposers.

15. (E)

In order to answer this question you must be familiar with the structure and functions of genes. Genes are DNA sequences that are found on chromosomes and code for polypeptides. A particular gene may exist in more than one form giving us alleles for the same gene. For example, two forms of the gene for pea plant height would be tall vs. short. When genes undergo change, they are said to mutate. Genes may be beneficial, neutral, or harmful to the survival of the organism. It is even possible for a gene to cause the death of the organism, in other words, to be lethal.

16. (C)

To answer this question you must apply your knowledge of the classification of mammals to a specific situation - - you must predict the characteristics that you would expect whales to have. Vertebrates, in general, have a vertebral column and dorsal hollow nerve cord. Since mammals are vertebrates, they would also have these characteristics. Birds and mammals are warm-blooded, and mammals nurse their young. Scales are a characteristic of fish and reptiles, and would not be expected on whales.

17. (D)

In order to interpret the results of this cross, we should first determine whether the gene for miniature wings is dominant or recessive. Then we can determine whether the gene is located on an autosomal, X, or Y-chromosome. Because the first generation of offspring is all wild-type, the gene for miniature wings must be recessive. If the gene were on the Y-chromosome, we would expect to find it passed from father to son every generation. However, it did not show up in the first generation of offspring. If the gene were on an autosomal chromosome, the 1/4 of the second-generation offspring that is miniature-winged would be expected to be both female and male. In X-linked inheritance, 1/4 of the second-generation offspring for this type of cross would be expected to show the recessive trait and, also, to be only males. Since the second generation of offspring showed 1/4 miniature-winged males, the trait must be due to a recessive X-linked gene.

Key: M = wild-type gene on X-chromosome
m = miniature wings gene on X-chromosome
Y = Y-chromosome

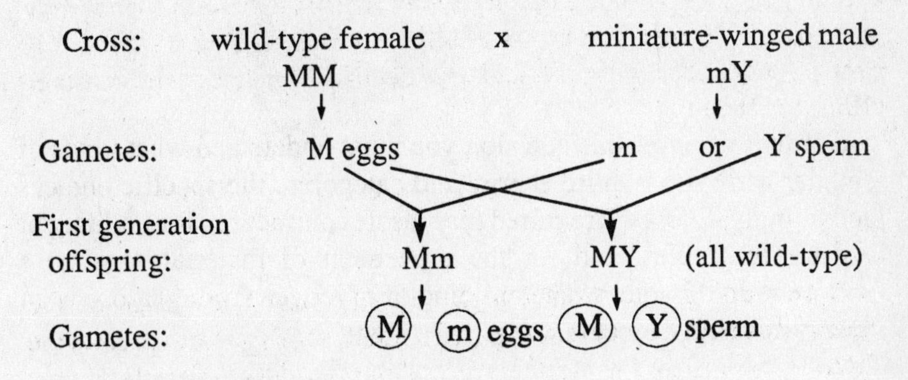

Punnet square:

	M	m
M	MM	Mm
Y	MY	mY

Second generation offspring:

genotypes:	**phenotypes:**
1/4 MM, 1/4 Mm	1/2 wild-type females
1/4 MY	1/4 wild-type males
1/4 mY	1/4 miniature-winged males

18. (C)

This question tests your ability to interpret experimental data given in graphical form. This graph shows the changes in abundance of two species over a period of time. Species A shows a sharp increase in number every 9 to 10 years, followed by a sharp decrease. Species B increases in number following the same cycle except that the maximum abundance is about one year later than that for species A. Since species B peaks after species A, it is most likely that species B is eating species A. When there are more species A organisms, there is more food for species B and B increases. Larger numbers of B feed on A and cause the decline of A. As A decreases, there is less food for B, and B dies off.

19. (D)

In order to answer this question you must understand what kinds of cellular activities require energy and categorize the specific choices accordingly. Energy is required for muscle contraction, for building of organic molecules, and for the movement of materials against a concentration gradient. The building of glycogen from glucose is an energy-requiring process.

The movement of a substance, such as sodium or potassium, across a cell membrane down a concentration gradient is diffusion. When materials move against a concentration gradient, they undergo active transport, an energy-requiring process. An example of this is the movement of calcium against a concentration gradient. Another example is the sodium-potassium pump that operates to maintain high concentrations of sodium outside and high concentrations of potassium inside the cell membrane.

Osmosis is the diffusion of water across a cell membrane. Since diffusion occurs along a concentration gradient, this is not an energy-requiring process.

20. (E)

This is a question that asks you to identify the site of meiosis. Meiosis is the process of cell division that reduces the number of chromosome sets to one. This occurs as part of sexual reproduction and results in haploid cells, which develop into gametes with half the number of chromosomes. When haploid gametes combine at fertilization, the chromosome number doubles and the zygote produced contains the normal diploid number of chromosomes. Therefore, the site of meiosis must be the testes in which sperm are made in preparation for fertilization.

Cell division in all other cells in the body is designed to produce daughter cells that are genetically identical to the parent cell. Therefore, cell division in stomach, skin, urethra, and intestine is by mitosis, and does not change the chromosome count.

21. (C)

To answer this question you must know the characteristics of an enzyme. Enzymes are specialized proteins made by living organisms that serve to speed up chemical reactions in the body. Thus, they serve as biological catalysts. They are specific in their actions reacting with particular substrates. Enzymes are not used up during the reaction, but rather can react over and over again with new molecules of substrate.

22. (A)

This is a question that asks you to apply the concept of niche to specific situations. A niche is the position or role that an organism occupies within an ecosystem. This includes its habitat, environmental factors (such as temperature tolerance, source of food), and behavior. If two organisms occupy the same niche, then they would be expected to have the same habitat, live under the same environmental conditions, eat the same food, and behave in the same general way. They would also be expected to mate during the same time.

23. (E)

The answer to this question depends upon your recognizing the fact that a peptide bond is the bond that forms between two amino acids. This bond forms because an -OH group is removed from the acid portion (-COOH) of one amino acid and a -H is removed from the amino portion (-NH_2) of the second amino acid. The -OH and -H combine to form H_20. This linking of amino acids is called a condensation reaction. (Note: R is different for each amino acid).

```
     amino acid 1                    amino acid 2

          R                               R
          |                               |
   H-N-C-C-(O-H)                   (H)-N-C-C-O-H
     |  |  ||                           |  |  ||
     H  H  O                            H  H  O

        R        R
        |        |
   H-N-C-C-N-C-C-O-H        +        H2O
   |   |  || |  |  |
   H   H  O  H  H  O
          ^
           \___ peptide bond
```

24. (D)

To answer this question you must understand why blood pressure drops as the blood moves away from the heart. Blood pressure is caused by the contraction of the ventricles of the heart. As the blood travels through the vessels, friction between the blood and the vessel walls slows it down. While the heart (right ventricle) does pump blood to the lungs first, the blood is returned to the heart and pumped out again from the left ventricle. Thus, the blood pressure measured in the body is due to the action of the left ventricle, and is not affected by circulation of blood to the lungs.

Answers (A), (C), and (E) are incorrect statements. Blood from the lower part of the body must return to the heart through veins against gravity. However, the blood from the upper body flows through veins with gravity. Arteries are not larger than their corresponding veins. Also, valves are not found in arteries. They are found in veins where they aid in the return of blood to the heart.

25. (C)

For this question you should be familiar with early development of humans. The ovum matures in the ovary and is released into the abdominal cavity from the ovary during ovulation. It is immediately picked up by the fallopian tube and begins to travel down the tube. If intercourse has occurred, sperm that were deposited in the vagina swim through the uterus and up the fallopian tube. Fertilization usually occurs in the upper portion of the fallopian tube. The zygote (fertilized egg) travels down the tube and begins to divide (cleavage). This dividing embryo enters the uterus and implants in the uterine wall.

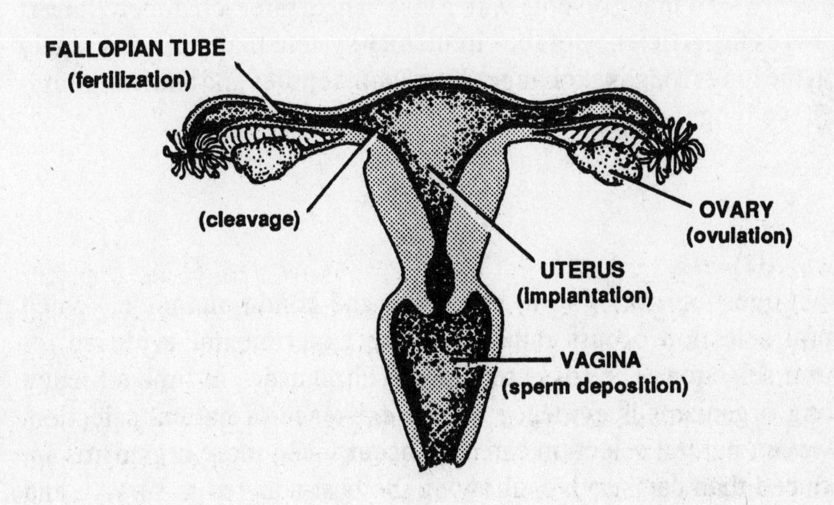

26. (A)

This question asks you to calculate the number of amino acids that would be present in a protein coded for by a bacterial gene containing a certain number of bases. Since the genetic code is a triplet code, a sequence of three bases codes for one amino acid. Therefore, 450 bases should code for 150 amino acids. This is an approximate number, since three different codes (UGA, UAG, and UAA) do not code for any amino acids. These are terminator codes, and one of them would be expected to end the reading of the gene.

27. (C)

This question is looking for a feature that is shared by amphibians and reptiles. Amphibians, the first land animals, have some characteristics that tie them to a moist environment. Eggs are laid and usually fertilized in a moist environment. While the adult forms have lungs, some respiratory exchange may occur across their thin, unprotected skin. Reptiles have developed several features that improve their ability to function as land animals. Their amniotic eggs contain a water supply and have shells that are impermeable to water. The amniotic egg requires that fertilization be internal rather than external. Reptiles have dry, scaly skin that prevents water loss. Their three-chambered heart improves the efficiency of the circulatory system that must rely solely upon the lungs for gas exchange. Thus, both reptiles and adult amphibians have lungs.

28. (D)

This question asks you to determine the conditions under which natural selection occurs rather than the experimental evidence for natural selection. The fossil record and similarities in limb structure among organisms is evidence for the existence of natural selection. However, natural selection can only occur when more organisms are produced than can survive, allowing the best-adapted to survive and reproduce. The fact that all living things contain DNA is another piece of evidence that organisms may have a common heritage, but is not a precondition for evolution. Natural selection may lead to extinction, but extinction does not cause natural selection.

29. (B)

This question deals with the problem of water loss in land plants, and asks you to figure out which mechanism would allow water loss to occur in the plant. The movement of water through the xylem occurs, because the water column is "pulled" up by evaporation of water from the leaf surface, a process called transpiration.

The water loss from transpiration must be balanced by water absorption by the roots. Therefore, the roots help to retain water in the plant. In addition, the leaf shape can reduce water loss by decreasing the surface area to volume ratio (as exemplified by round pine needles). The impermeable waxy cuticle that covers plant parts exposed to the air also prevents water loss. Stomata are the gas exchange openings in leaves plus their two guard cells. Because guard cells can close when photosynthesis is not occurring and when water loss would be severe (during the heat of the day), this is an important mechanism for water retention in the plant.

30. (D)

For this question you need to know how protein synthesis occurs. The nucleus contains genes on chromosomes, and the genes direct the synthesis of RNA. Therefore, the nucleus is the site of the RNA production. Messenger RNA is a copy of a gene that codes for a particular protein. It travels from the nucleus to the ribosome where it attaches and directs the assembly of the protein. Thus, translation occurs at the ribosome. Transfer RNA transports the amino acids to the ribosome for protein synthesis.

31. (C)

This question asks you to recognize the ecosystem that is dominated by evergreen trees. This is the taiga, an area of extremely cold winters. The tundra, a region north of the taiga, has a permanently frozen subsoil (permafrost) and cannot support any large plants. The temperate forests are farther south and contain deciduous plants (plants that lose their leaves) that need more rainfall and a relatively long, warm summer. Tropical rain forests have abundant rainfall and are characterized by tremendous species diversity, with up to ten dominant trees. A chaparral is a region dominated by drought-resistant and fire-resistant small trees and shrubs, because of the cool, rainy winters and hot dry summers.

32. (E)

To answer this question you must know how evolution occurs, and must figure out whether specific situations lend themselves to evolutionary change. In order for evolutionary change to occur, organisms must exist that are genetically different from each other. Acquired changes in the individual will not contribute to evolution, because there is no genetic basis for these changes. Even if the acquired changes increase reproductive success, the offspring will not show an increase in the frequency of the acquired trait.

Thus, a nutritionally well-balanced diet may improve the horses that are given such a diet, but will not change the next generation horses. On the other hand, genetically different forms of birds provide the raw materials for evolution. Those organisms that are genetically better-adapted to the environment, will be the most likely to survive and reproduce. For example, antibiotic-resistant bacteria will be selected for after exposure to antibiotics. Behavioral characteristics that are not acquired would be inherited and could be selected either during natural selection or selective breeding.

Another source of evolutionary change is genetic drift. Given genetic variation in a small population, it is possible that some organisms may survive based upon luck. If the individuals surviving a hurricane happen to be genetically different from the rest of the population, then evolutionary change will occur. Genetic drift does not necessarily mean that the surviving organisms are the best adapted or most fit.

33. (B)

This question asks you to identify the limiting factor(s) in the growth of trees in a grassland ecosystem. To answer this question you must know that grasslands are a climax community, in which periodic droughts and fires prevent the larger water-dependent trees from surviving. Grasses survive droughts by becoming dormant and survive fires by reseeding themselves.

34. (D)
To answer this question you must use your knowledge of the chemical structures of proteins and DNA to find what is different about them. Proteins are made of twenty different amino acids with the following general structure:

```
        R
        |
H–N–C–C–O–H
  |  |  |
  H  H  O
```

R is the group that is different for each amino acid. All amino acids contain carbon, hydrogen, nitrogen and oxygen, but not phosphorus.

DNA is made of nitrogen containing bases, a sugar (deoxyribose), and phosphates. The nitrogen-containing bases and sugar are organic compounds that also contain carbon, hydrogen, and oxygen. The phosphates are compounds containing phosphorus and oxygen. Therefore, DNA has all of the elements listed as choices. The fact that DNA contains phosphorus while protein does not is a way of distinguishing between the two chemically.

35. (D)
This is a straightforward question about the structure of bacteria. Bacteria are prokaryotes, and, therefore, have their genetic material organized as a large circular molecule of DNA. They do not have distinct nuclei, chloroplasts, or mitochondria. In addition, their cell membranes are surrounded by outer cell walls.

36. (C)

This question requires that you know something about the autonomic nervous system, and be able to figure out what types of activities might be autonomic in function. The autonomic nervous system is the part of the peripheral nervous system that includes the motor neurons that are not under voluntary control. These neurons control smooth muscle (found in internal organs), the heart, and glands. Thus, the autonomic nervous system controls the smooth muscle of the digestive tract, heart rate, and salivary as well as adrenal gland secretions. Contraction of skeletal muscles is under the control of the somatic nervous system.

37. (C)

This question asks you to distinguish between fungi and plants. Both fungi and plants are eukaryotes, may be multicellular, produce spores during parts of their life cycles, and have cell walls. Plants undergo photosynthesis, while fungi obtain energy and raw materials for growth by decomposing living or dead organic material.

38. (D)

To answer this question you must be familiar with how various organisms reproduce. While yeast, hydra, bacteria, and *Paramecium* can undergo sexual reproduction, they also have the ability to reproduce asexually. Budding occurs in yeast and hydra, while binary fission occurs in bacteria and *Paramecium*. Birds can only reproduce by sexual reproduction, the union of genetic material from two different parents.

39. (B)

To answer this question you must be able to solve a genetic problem. To do this you must determine the genotypes of the parents, the gametes produced, the genotypes of the offspring, and tabulate the phenotypes of the offspring. Let's begin with the gene symbols:

Key: R = round seed (dominant)
r = wrinkled seed (recessive)

Y = yellow (dominant)
y = green (recessive)

Cross: RrYy x rrYy

Gametes: Because the genes are on separate chromosomes, they assort independently of each other.

Therefore, RrYy- - - - RY, Ry, rY, ry gametes

And, rrYy- - - - rY, ry gametes

Punnet square:

	RY	Ry	rY	ry
rY	RrYY	RrYy	rrYY	rrYy
ry	RrYy	Rryy	rrYy	rryy

Phenotypes:

3/8 RrY- (round yellow seeds) 1/8 Rryy (round green seeds)
3/8 rrY- (wrinkled yellow seeds) 1/8 rryy (wrinkled green seeds)

Therefore, the phenotypic ratio is 3:3:1:1.

40. (A)
To answer this question you need to recognize the different types of tissues in plants. Meristematic tissue is the undifferentiated, actively dividing tissue that is responsible for growth. Therefore, the meristem

would show the greatest numbers of miotic divisions. Cork (thick, water-impermeable cells in bark) and xylem (hollow vessels that conduct water and minerals from roots to leaves) are both made of dead cells. Phloem and guard cells are highly differentiated. Phloem transports sap and guard cells regulate the size of stomata.

41. (E)

To answer this question you must figure out what happens when parents devote resources and energy to the rearing of the young. It might be helpful to first look at the situation in which there is no parental involvement.

Organisms that produce large amounts of offspring depend upon the sheer size of the litter for the perpetuation of their species. The young mature very quickly, and are not educated, as the parents are usually involved with obtaining their own food and with reproduction. Should some of the offspring become endangered, the parents will not interfere, because it is not expected that all the young survive, which is the reason for a large litter.

Organisms that are much involved in the education of their young would place the welfare of their young above their own because such organisms usually have few offspring, as few as one per season (e.g.: humans), and the death of even one of the young would seriously jeopardize the chances of any offspring surviving to maturity. In such cases, therefore, survival to maturity (of any of the young) depends upon education of the young by the parents (this means that more time is necessary in order for the parents to impart their knowledge to their young), smaller litters (it would be impossible for two parents to thoroughly educate too many offspring), and, if necessary, protection of the young (the parents have already survived to reproduce; the young have not).

42. (E)

To answer this question you must know what a gene is and how mutations can affect the function of the gene. A gene is a sequence of nucleotide bases that codes for the amino acid sequence of a protein.

The code is a triplet code, and three bases code for each amino acid. One type of mutation that does not change the length of the gene is a change in a single base.

Let's look at what would happen to the protein if there were a single base change. Since many of the amino acids have more than one code, a change in the triplet code might produce an alternate code for the same amino acid. Therefore, the protein might not change at all. If the mutation does result in an amino acid change, it might not be in a critical portion of the protein. The base substitution could be beneficial, if the structure of the protein improves. It could have no effect or it could be harmful.

43. (D)

This question asks you to remember the characteristics of protozoa. They are eukaryotes and, therefore, have the typical membrane-bound internal structures. They are primarily single-celled organisms that must ingest food to survive. The different groups of protozoa are characterized by the variety of techniques for locomotion: Mastigophora use flagella, Sarcodina use pseudopods, Ciliata use cilia, while Sporozoa lack special structures for locomotion.

44. (D)

It is important to know what diploid means and you must know what cells are diploid. Diploid refers to a condition of having two sets of chromosomes. Spores, eggs, and sperm are all formed by meiosis, a process of cell division whose purpose is to reduce the amount of chromosomes to one set. Therefore, spores, eggs, and sperm are haploid. Drones develop from unfertilized eggs and, therefore, are also haploid. Zygotes are products of eggs fusing with sperm and are, thus, diploid.

45. (C)

To answer this question you must know something about ecological pyramids. Energy flow, numbers of organisms, and biomass pyramids may be constructed for a particular ecosystem. These show the primary producers (plants) at the bottom, then primary consumers (herbivores), then secondary consumers, and finally tertiary consumers at the top. Members of each level feed on the members of the level below.

Productivity (photosynthesis) is only measured at the first level, since that is where the plants are. As one moves up the pyramid, the energy levels decrease, because energy transfers between trophic levels are inefficient and result in heat loss to the environment. Since energy levels decrease, usually the biomass would also decrease and the number of organisms would also decrease (less energy supports fewer organisms). The exception to the biomass pyramid occurs when the producers have a very high rate of reproduction. The exception to the population pyramid occurs when the primary producer is very large (for example, a tree).

46. (A)

To answer this question you must know how chemical reactions work in the living organism and apply that knowledge to a specific situation. The rate of chemical reactions increases with an increase in temperature. This typically happens with poikilotherms (cold-blooded animals) as the environmental temperature increases, but also happens during fever. Since cellular respiration is a series of chemical reactions, the rate of cellular respiration will also increase, resulting in greater energy production. Thus, energy-dependent activities will not be limited by an increase in body temperature. Temperature changes should not affect the formation of enzymes.

47. (B)

To answer this question you must know how chromosomes behave during meiosis. Chromosomes contain genes linked together in a linear fashion. Genes on the same chromosome are usually inherited together, unless crossing over occurs. Crossing over can occur during prophase I of meiosis, because homologous (like) chromosomes, each containing their replicated (chromatids) strands, line up next to each other. During crossing over, one chromatid from each homologous chromosome breaks and reunites so that the new chromosomes contain pieces of the original chromosomes.

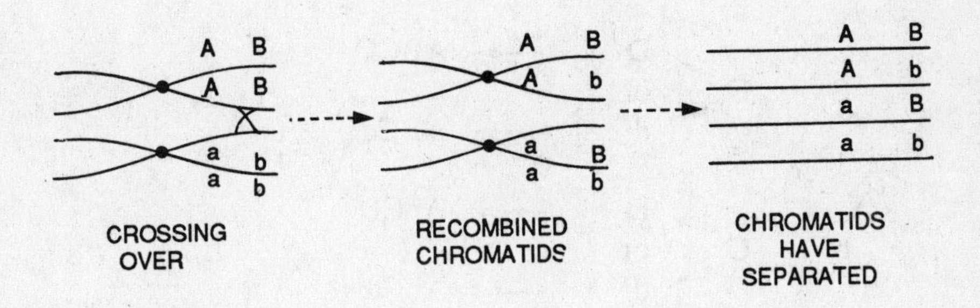

If the genes are on different chromosomes they will assort independently of each other. The frequency of chromosomal rearrangements, such as translocations, deletions, duplication, and inversions, is not related to gene linkage. Blending inheritance occurs in some situations where the heterozygous condition shows an intermediate phenotype between an individual who is homozygous dominant and one who is homozygous recessive.

48. (A)
This question tests your ability to draw conclusions from experimental results shown in a graph. The graph shows the number of mutations that occur after exposure to different concentrations of a variety of chemicals. Different chemicals cause different numbers of mutations.

As the amount of chemical increases the number of mutations also increases. In some cases (chemicals C, D, E) high concentrations of a chemical result in only a minor increase in mutations. Chemical A clearly causes the most mutations. This graph does not tell us whether the mutations are harmful, beneficial, or neutral.

49. (E)

To answer this question you must figure out the different kinds of gametes that can be produced.

```
                    ↗ D    ----► ABCD
      B  →  C         or
     ↗              ↘ d    ----► ABCd
   A    or
     ↘              ↗ D    ----► AbCD
      b  →  C         or
                    ↘ d    ----► AbCd
  or

                    ↗ D    ----► aBCD
      B  →  C         or
     ↗              ↘ d    ----► aBCd
   a    or
     ↘              ↗ D    ----► abCD
      b  →  C         or
                    ↘ d    ----► abCd
```

There are eight different gametes possible. An easier way of doing this is to see that there are two choices for A(A or a), two for B (B or b), one for C, and two for D (D or d). Therefore, 2x2x1x2 equals eight.

50. (B)

This question can be answered with an understanding of the structure of the seed. The seed consists of an embryo, stored food that is endosperm or formed from endosperm, and the outer protective layers of the ovule or seed coat. As the embryo develops, one portion (the radicle) becomes the root and another portion (the epicotyl) becomes the stem. In angiosperms, the ovules are found within the ovary, which develops into the fruit. The fruit protects the seeds from water loss during development and may aid in dispersal of the seeds later.

51. (A)

To answer this question you need to know the functions of the three different forms of behavior. Social behavior involves the interactions of organisms within a species and includes, among others, mating behavior, caring for young, food procurement, and protection from predators. These all depend upon a high degree of communication. Echolocation is a mechanism used by bats for finding their way when they are flying. They emit sound waves that bounce off objects and these sound waves are, in turn, perceived by the bats. Circadian rhythms are regular cycles of activity that occur about every 24 hours. Neither echolocation nor circadian rhythms depend upon interaction with other organisms, and, therefore, they do not depend upon social communication.

52. (D)

For this question you must recognize the different sources of genetic variability in a population. These include changes in the base sequence of a gene or in chromosomal organization which are called mutations. The chromosomal rearrangements might be translocations, duplications, deletions, or inversions. Migration of genetically different organisms into a population would, also, increase genetic variability. Sometimes a particular phenotype depends upon the fact that certain genes are inherited together. Recombination, either by independent assortment of different chromosomes or crossing over of homologous chromosomes, would give new combinations of genes that might alter the phenotype of the next generation.

Once genetic variation exists, natural selection acts to choose those individuals that are best adapted to the particular environment. Thus, natural selection is not a cause of genetic variability, but rather, a possible consequence of it.

53. (B)

To answer this question you must be familiar with ecological relationships among organisms, in particular predator-prey relationships and how mimicry protects the prey. If an organism has an unpleasant taste, a predator that has experienced the taste will avoid eating that organism. Another organism that looks like the first will also be protected, because it will be mistaken for the first. Therefore, mimicry of species B with the unpleasant taste will protect species A, because species A looks like species B.

54. (E)

To answer this question you must be familiar with subcellular organization. Eukaryotic cilia and flagella are external structures that are anchored to the cell's basal body by microtubules. In single-celled organisms, they are responsible for locomotion. Cilia found in multicellular organisms move substances across the cell surface. Spindle fibers connect chromosomes to their respective poles during cell division. They contain microtubules that lengthen and shorten, first to move the chromosomes to the center of the cell during metaphase, then to move chromosomes to the poles. Actin and myosin are components of muscle cells that react with each other to cause muscle contraction. Lysosomes are not involved in movement, but store potent digestive enzymes that can be used for the breakdown of cellular components.

55. (B)

This question asks about ecological relationships among organisms that define the source of their food. Foxes are predators, since they attack and eat their prey, the rabbits. Organisms that live in or on, and feed on other living organisms are called parasites. Predators eat the prey when it is dead. Producers are plants and make their own organic materials. Herbivores eat plants. Decomposers feed on dead organic matter.

56. (D)

This question tests your understanding of what happens in a population that is not stable. (This is a population that does not follow the Hardy-Weinberg law). If the heterozygotes have a reproductive advantage, this means that there will be more offspring produced from a mating between heterozygotes. Since the offspring of such a mating are expected to be 1/4 homozygous dominant, 1/2 heterozygous, and 1/4 homozygous recessive, this ensures the fact that dominant and recessive homozygotes will be maintained in the population even if they are not as reproductively successful.

57. (E)

This question requires some knowledge of the characteristics of angiosperms as compared with those of other plants. Tracheophytes have vascular tissue and have true roots and leaves. The seed-producing plants (gymnosperms and angiosperms) have a gametophyte that is dependent upon the sporophyte. Only angiosperms produce flowers and fruit.

58. (C)

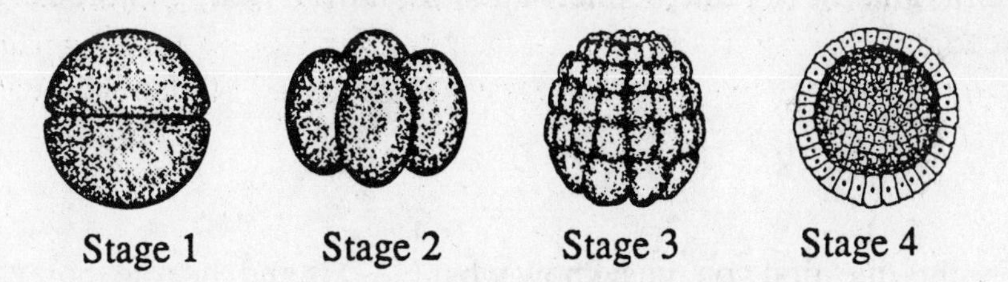

To answer this question you must recognize the stages in embryonic development and predict the next stage. The first three drawings show early cleavage leading to the formation of a ball of cells called the morula. This ball of cells hollows out to become the blastula, shown as stage four. The next stage is gastrulation. One end of the ball invaginates and the cells move inward to produce a two-layered structure. Later, the dorsal surface will invaginate during formation of the neural tube, the forerunner of the spinal cord.

Questions 59 - 62

This group of questions deals with the functions of various cellular structures and in particular those functions that relate to cellular respiration and photosynthesis.

59. (C)

To answer this question you must know what happens during glycolysis and where it occurs. Glycolysis is the first part of cellular respiration. Glucose is broken down to pyruvic acid during aerobic respiration or to lactic acid during anaerobic respiration. Glycolysis occurs in the cytoplasm.

60. (E)

This question asks you to identify the structure that stores water as well as glucose. Water is taken up by plant vacuoles. Because plant cells have cell walls, the hypertonic cell can fill with water to maintain turgor pressure without bursting the cell. In addition, other materials such as glucose (an end product of photosynthesis) may be stored in the vacuole.

61. (A)

For this question you must know what NAD is and the role it plays in cellular respiration. NAD (nicotinamide adenine dinucleotide) accepts hydrogens and electrons to form NADH during cellular respiration. The NADH produced during glycolysis travels into the mitochondrion, while NADH produced during the Krebs cycle remains in the mitochondrion. Inside the mitochondrion, NADH transfers its electrons and hydrogen to the electron transport chain. The transfer of these electrons down the chain releases energy that is used to make ATP.

62. (D)

This question assumes that you know that plants are the only organisms that can convert light energy to chemical energy. Chloroplasts contain chlorophyll and other pigments that have the ability to capture light energy, raising the energy level of certain chlorophyll electrons. As the excited electron returns to its normal energy level, the energy is converted to chemical energy.

Questions 63 - 65

This set of questions concerns the role of the various flower parts in the processes of pollination and fertilization.

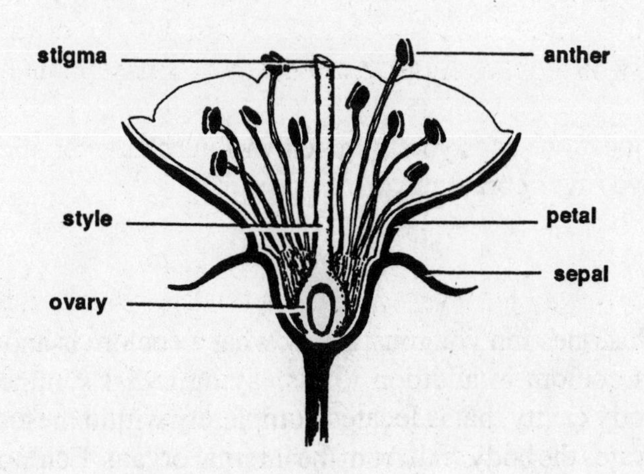

63. (A)

To answer this question you must know that pollen contains two sperm nuclei and possibly a tube nucleus. Therefore, pollen is formed on the male part of the flower called the anther.

64. (D)

This question assumes you know what pollination is. This is the transfer of pollen grains from the anther to the stigma of the female flower part. When cross-pollination occurs (pollination between separate flowers) the actions of wind or insects become important. Insects may be attracted to a flower by its brightly-colored petals as well as its aroma. The sepals are the outermost protective leaf-like parts of the flower. They are usually green and do not serve to attract insects.

65. (C)

The answer to this question depends upon your knowing that the egg nucleus is located in the ovary. After the pollen grain lands on the stigma, the tube nucleus forms a pollen tube through the style to the ovary. Within the ovary, a sperm nucleus fuses with the egg nucleus to form the zygote. (Another sperm nucleus fuses with two other nuclei in the ovary to form the endosperm).

Questions 66 - 68

This set of questions asks you to match various characteristics with their respective invertebrate phyla.

66. (A)

To answer this question you must know what a coelom is and which phylum has a coelom in addition to displaying radial symmetry. A coelom is a body cavity that is located completely within mesodermal tissue. It separates the body wall from the internal organs. Echinoderms have a coelom, but cnidarians do not. Cnidarians have a mesoderm that is sandwiched between an ectoderm and endoderm, with the only cavity being inside the endoderm (gastrovascular cavity). Both cnidarians and adult echinoderms have radial symmetry.

67. (D)

For this question you must know what is meant by segmented and which organisms are segmented worms. When an organism is segmented, its body is divided into roughly equal units containing repeating internal and external structures. Platyhelminthes are flatworms and do not have repeating structures. However, annelids are worms whose repeating segments contain structures such as a separate coelom, nephridia, and setae.

68. (E)

For this question you must be familiar with the types of skeletons found in different organisms. Cnidarians, platyhelminthes, and annelids have hydrostatic skeletons. This type of skeleton depends upon alternating muscles in the body wall contracting and putting pressure on the rest of the body. Echinoderms have an endoskeleton, a skeleton made of plates embedded in the skin of the organism. An external skeleton with jointed appendages for movement is found in arthropods.

Questions 69 - 70

This group of questions concerns the functions of structures that are important in motion.

69. (A)

For this question you must be familiar with the way in which a joint is held together. Muscles connect to the bones of a joint and move the joint when they contract. Muscles are connected to bone by thick strands of dense connective tissue called tendons. The articulating surfaces of the bones are covered by cartilage. In addition, the joint is stabilized by ligaments which connect bones to each other.

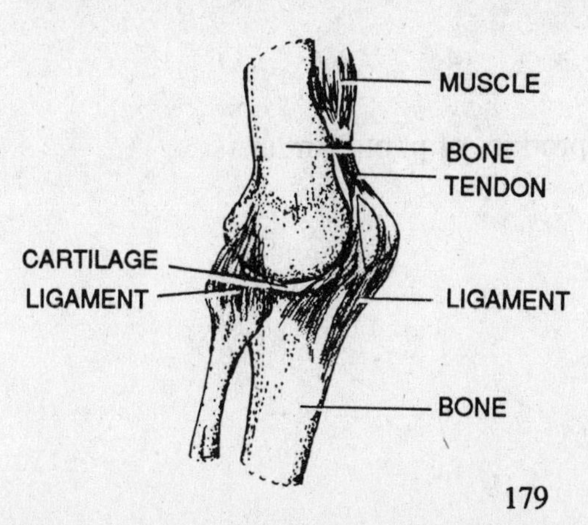

70. (D)

To answer this question you must realize that bones contain living cells that must receive oxygen and nutrients to survive. As with all other cells of the body, these essential materials are delivered by the circulatory system.

Questions 71 - 74

These questions test your knowledge of chemical reactions that occur in plants, mostly referring to photosynthesis.

71. (E)

To answer this question you must know what happens during photosynthesis. Light energy captured by the plant is converted to chemical bond energy in the form of the energy intermediates, ATP and NADPH. The energy from ATP and NADPH is used to produce glucose from carbon dioxide.

72. (D)

It is important that you know what happens to the electrons in chlorophyll that capture the light energy. These excited electrons are passed along to NADP to make NADPH. The missing electrons in chlorophyll are then replaced by electrons from water. The water molecule splits to form hydrogen ions (H^+), free electrons for transfer to chlorophyll, and oxygen. It takes two water molecules to make one molecule of oxygen.

$$2H_2O \longrightarrow 4H^+ \quad + \quad 4e^- \quad + \quad O_2$$

Therefore, oxygen is an end-product of photosynthesis.

73. (A)

For this question you must be familiar with the chemical structure of the plant cell walls. Glucose combines to form long chains called cellulose. This cellulose is the basic component of plant cell walls.

74. (C)

For this question it is important that you recognize that cells must have energy available in a readily usable form. ATP is formed when ADP and a phosphate combine. The phosphate can attach and break off easily. Energy is required for ATP production, which occurs during photosynthesis in plants. In animal cells, ATP forms during cellular respiration. When ATP breaks down to ADP and phosphate, energy is released and available for other cellular functions. ATP is, therefore, the energy currency of plant and other cells.

$$\text{ADP} + \text{phosphate} \underset{\text{energy given off}}{\overset{\text{energy needed}}{\rightleftharpoons}} \text{ATP}$$

Questions 75 - 78

This set of questions tests your knowledge of the functions of glandular structures associated with digestive activity.

75. (A)

To answer this question you must be familiar with the first steps in digestion. The salivary glands make an enzyme called salivary amylase that mixes with food in the mouth and begins the digestion of starch to form glucose. The high acidity of the stomach inactivates this enzyme before it has a chance to complete digestion.

76. (C)

For this question you must know which digestive enzymes are produced by the various organs in the digestive system. Pepsin is produced in the stomach and digests protein.

77. (D)

To answer this question you must realize that bile is made in one organ, but stored in another. The liver makes bile, which is then stored in the gall bladder until needed for fat digestion.

78. (E)

To answer this question you must be familiar with the accessory organs that aid in digestion. The pancreas is not only an endocrine gland that makes insulin, but it also makes digestive enzymes that act on food in the small intestine. Insulin is the hormone that increases membrane permeability to glucose and, therefore, regulates blood sugar levels. The digestive enzymes of the pancreas help break down carbohydrates, fats, and proteins.

Questions 79 - 83

This set of questions refers to the diagram of a reflex arc, in this case the knee jerk. To answer these questions, you must identify the structures in the diagram and associate them with the functions they serve in a reflex.

79. (C)

To answer this question correctly you must identify the sensory neuron and find its receptor. The sensory neuron is labeled 4. An impulse begins at the sensory receptor (located at 3) in the muscle, and travels the length of the neuron to the spinal cord, ending at 6. The sensory receptor is stimulated when the muscle lengthens.

80. (B)

For this question you must realize that the sensory neuron has a very long dendrite. Its cell body, which contains the nucleus, is located next to the spinal cord at 5. The group of cell bodies of all the sensory neurons entering the spinal cord at this point is called the dorsal root ganglion.

81. (C)

To answer this question you must know what a synapse is. This is the space between two neurons, the terminal branches of the first neuron and the dendrites of the second neuron. More than one neuron can synapse on another single neuron, which means that information from different areas of the body can be integrated; an appropriate response can then be effected. A specialized synapse called a myoneural junction occurs between a neuron and the muscle cells it stimulates. An example of a myoneural junction is 9, but this was not one of the choices for this question.

82. (D)

This question assumes you know the structure of the spinal cord. It contains an inner area of grey matter, which contains the cell bodies of motor neurons and interneurons. The outer area of white matter contains tracts that go to or come from the brain. Therefore, information that travels to and from the brain must travel along 7. It must be noted, however, that in the actual reflex arc, no contact with the brain is made with respect to the actual motor response. Contact with the brain is made by sensory neurons that relay the information concerning the actual movement of the leg.

83. (E)

This question asks you to identify the myoneural junction, the synapse between the motor neuron and the muscle. This synapse consists of the terminal branches of the motor neuron and the muscle cell immediately below it and is located at 9.

Questions 84 - 86

For this set of questions you are asked to analyze the results of a series of experiments showing the effects of two plant hormones on growth. It is necessary to read and interpret the data as presented in graph form.

84. (E)

For this question you must be able to look at all four graphs and figure out which one shows the most plant growth. Plant cells in experiment IV show the greatest growth. This experiment involves 25 mg/l auxin and different concentrations of kinetin. The greatest growth shown for this experiment occurred when kinetin had a concentration of about 1 mg/l.

85. (B)

To answer this question you must know that a control is the experimental situation that leaves out the variable. The variable, in this case, is the amount of auxin present. Therefore, the control is experiment I, because no auxin is present in this experiment.

86. (A)

For this question you must draw a conclusion about the results shown. The greatest growth occurred when both hormones were present together, making A the correct answer.

When auxin was not present, high concentrations of kinetin increased growth slightly. On the other hand, high concentrations of kinetin, in the presence of auxin, inhibited growth in experiments II, III, and IV.

Growth occurred as long as one of the hormones was present. In experiment I, growth occurred in the presence of kinetin even though auxin was missing. In experiments II, III, and IV growth occurred at 0 mg kinetin, because auxin was present.

Auxin affected growth more than kinetin. To see the effects of auxin alone, you must look at growth of 0 mg kinetin in experiments II, III, and IV. Compare these readings with the effects of kinetin alone, shown in experiment I.

Questions 87 - 88

To answer this set of questions you must know the chemical structure of the cell membrane and what it does.

87. (B)

For this question you must be familiar with the chemical structure of the membrane. The membrane is primarily a double layer of phospholipid molecules. These phospholipids are composed of phosphate ends that are oriented to the surfaces of the membrane, as well as fatty acid chains pointing to the inside of the membrane. Interspersed between the phospholopids are proteins. Some of these proteins extend to the outside of the membrane, while others are located within the membrane. The proteins serve a variety of functions, including acting as receptors on the surface and serving as pores through which certain materials can pass into and out of the cell.

88. (A)

For this question an understanding of the structure of the membrane will help you predict its behavior. Because the membrane is made primarily of lipids with uncharged fatty acid chains, charged particles, such as salts, are repelled by the lipid portion. Instead, the protein pores allow for passage of these particles. In addition, large particles, such as proteins including proteineous hormones, have difficulty passing through the membrane. However, they may have an effect on the cell by binding with protein receptors on the outside of the cell. The cell is not permeable to proteins, because proteins are too large and may have charged portions.

The membrane has a positive charge on the outside, because of an unequal distribution of certain ions. Positive sodium ions are found in higher concentration outside the membrane, while organic materials found inside the cell are more likely to have a negative charge. The maintenance of this concentration of sodium ions outside depends upon an expenditure of energy. This process, called active transport, also functions in moving other materials across the membrane against their concentration gradients.

Questions 89 - 92

For this group of questions you are asked to analyze data in table form related to the different fluid compartments of the body.

89. (C)

This question asks you to compare the amounts of materials present in the different fluids. While urine only contains three of the five materials listed, it shows the greatest variability in the amounts of the materials present. Sodium and potassium amounts vary. This makes sense, since the function of urine is to rid the body of waste materials or materials that are present in excess. It is important for the internal body conditions to remain relatively constant, if it is to function properly. Materials, such as sodium and potassium, may vary depending upon dietary intake and metabolic needs.

90. (A)

For this question, you must draw a valid conclusion by analyzing the data on the table. The largest component in lymph is protein, not glucose. The largest component in urine is urea; normally no protein is found in urine. When cerebrospinal fluid and blood are compared, they have the same components, but in very different concentrations. As you can see, glucose is normally not excreted by the body, since no glucose is found in urine.

When we compare lymph, blood, and intercellular fluid, they contain the same amounts of glucose, urea, sodium, and potassium. They do differ in the amount of protein present. The capillaries of the circulatory system are impermeable to proteins. However, slight amounts of protein leak out and are quickly picked up by the highly protein-permeable lymphatic vessels.

91. (B)
For this question you must be familiar with the general composition of lymph. In addition to the components listed, lymph also circulates white blood cells. Red blood cells, albumin, platelets, and fibrinogen are found in the circulatory system, but are not present in lymph.

92. (B)
To answer this question you must know what nitrogenous wastes are and how they are removed from the body. When proteins are broken down, the liver converts nitrogen-containing amino groups to ammonia. Ammonia, however, is a toxic material, and must be converted to a nontoxic substance. In mammals the ammonia is then converted to the less toxic urea prior to being stored in the bladder, from which it exits the body.

Questions 93 - 94

This set of questions requires that you understand how an experiment dealing with nuclear transplants is designed and draw a valid conclusion from the experiment.

93. (A)

For this question you must know for what the experiment is testing and, from that, figure out what the appropriate control should be. These transplants are testing the effect of nuclei from various sources on the development of an egg. It is possible that some cytoplasm from the donor cell may be transferred with the nucleus, and it is, also, possible that the process of injection may trigger development changes. To determine the effects of cytoplasm and of the injection on the egg cell (so that if such effects occur in the experimental ova, they can be attributed to the presence of cytoplasm or the act of injection) injecting cytoplasm would serve as a control.

94. (A)

To answer this question, you must look at what the experiment is testing. It is designed to see whether the various nuclei can control development of the egg. Blastula, gastrula, and intestinal nuclei can support normal development, while cytoplasm alone can not. However, since fewer eggs containing gastrula and intestinal nuclei develop normally, changes must have occurred in the nucleus during development. This experiment is not testing immune reactions or looking for genes that are lost or mutated.

Questions 95 - 96

For this set of questions you must be familiar with the inheritance of the gene for sickle cell anemia.

95. (C)

This question asks you to interpret the data given in the table concerning the chemical makeup of hemoglobin. Remembering that a particular gene codes for one protein (polypeptide), if only one form exists, the homologous genes must be the same. Likewise, if two forms of hemoglobin are present, then two forms of the gene for hemoglobin must also be present. Therefore, individuals A and B are homozygous for the hemoglobin gene, while individual C is heterozygous. Thus, individual B does not carry any normal hemoglobin gene.

To determine dominance we must look at the heterozygote. However, in this case the question of dominance depends upon whether we are looking at the presence of disease or the chemistry of the individual. The heterozygote is phenotypically normal, and, therefore, the gene for sickle cell anemia would be classified as a recessive gene. However, the heterozygote contains both types of hemoglobin, and, therefore, shows codominance for the hemoglobin gene. Many other genes show typical dominant-recessive behavior when looking at the physical appearance of the offspring, but codominance or intermediate inheritance when we look at the chemistry of the gene.

96. (C)

To answer this question you must know how alleles (forms of the same gene) generally differ. First of all, a gene is made of DNA rather than RNA, and, therefore, would not contain uracil or ribose. A different form of a gene might mean that the gene is missing or that there is a base substitution. Since a different form of hemoglobin is found when the defective gene is present, the gene must be present but in altered form.

Questions 97 - 100

To answer this set of questions you must read and analyze results presented in graphical form showing changes in a population of algae and changes in environmental variables.

97. (C)

This question asks you to interpret the data shown on the graph. You must look at the unbroken line showing the number of algae present and find the steepest portions slanting upward. There are marked increases in growth in both the spring and autumn.

98. (D)

To answer this question you must compare the levels of nutrients, light, and temperature during the summer, yet algal peaks do not occur then. On the other hand, nutrient levels are low and must, therefore, be the limiting factors.

99. (E)

To answer this question you must compare the changes in environmental factors with peaks in algal growth. In the winter, nutrient levels are high, but light and temperature are at their lowest. Therefore, light and temperature must be limiting factors then. Maximum growth in the spring corresponds to high nutrient levels with increasing amounts of light and temperature. As the algae grow they use up nutrients in the environment, which limits their continued growth. In the fall, nutrient levels begin to increase again. Growth increases and is sustained while temperature and light levels are still high.

100. (E)

To answer this question you must know what nutrients are needed by algae and how they are supplied. Carbon dioxide is needed for photosynthesis and is available dissolved in the water. In addition, a variety of other nutrients are needed: nitrogen for protein synthesis, phosphorus for DNA, and others. These materials are most easily supplied from dead organic matter when liberated by decomposers.

BIOLOGY

TEST IV

BIOLOGY
ANSWER SHEET

1. (A) (B) (C) (D) (E)
2. (A) (B) (C) (D) (E)
3. (A) (B) (C) (D) (E)
4. (A) (B) (C) (D) (E)
5. (A) (B) (C) (D) (E)
6. (A) (B) (C) (D) (E)
7. (A) (B) (C) (D) (E)
8. (A) (B) (C) (D) (E)
9. (A) (B) (C) (D) (E)
10. (A) (B) (C) (D) (E)
11. (A) (B) (C) (D) (E)
12. (A) (B) (C) (D) (E)
13. (A) (B) (C) (D) (E)
14. (A) (B) (C) (D) (E)
15. (A) (B) (C) (D) (E)
16. (A) (B) (C) (D) (E)
17. (A) (B) (C) (D) (E)
18. (A) (B) (C) (D) (E)
19. (A) (B) (C) (D) (E)
20. (A) (B) (C) (D) (E)
21. (A) (B) (C) (D) (E)
22. (A) (B) (C) (D) (E)
23. (A) (B) (C) (D) (E)
24. (A) (B) (C) (D) (E)
25. (A) (B) (C) (D) (E)
26. (A) (B) (C) (D) (E)
27. (A) (B) (C) (D) (E)
28. (A) (B) (C) (D) (E)
29. (A) (B) (C) (D) (E)
30. (A) (B) (C) (D) (E)
31. (A) (B) (C) (D) (E)
32. (A) (B) (C) (D) (E)
33. (A) (B) (C) (D) (E)
34. (A) (B) (C) (D) (E)
35. (A) (B) (C) (D) (E)
36. (A) (B) (C) (D) (E)
37. (A) (B) (C) (D) (E)
38. (A) (B) (C) (D) (E)
39. (A) (B) (C) (D) (E)
40. (A) (B) (C) (D) (E)
41. (A) (B) (C) (D) (E)
42. (A) (B) (C) (D) (E)
43. (A) (B) (C) (D) (E)
44. (A) (B) (C) (D) (E)
45. (A) (B) (C) (D) (E)
46. (A) (B) (C) (D) (E)
47. (A) (B) (C) (D) (E)
48. (A) (B) (C) (D) (E)
49. (A) (B) (C) (D) (E)
50. (A) (B) (C) (D) (E)
51. (A) (B) (C) (D) (E)
52. (A) (B) (C) (D) (E)
53. (A) (B) (C) (D) (E)
54. (A) (B) (C) (D) (E)
55. (A) (B) (C) (D) (E)
56. (A) (B) (C) (D) (E)
57. (A) (B) (C) (D) (E)
58. (A) (B) (C) (D) (E)
59. (A) (B) (C) (D) (E)
60. (A) (B) (C) (D) (E)
61. (A) (B) (C) (D) (E)
62. (A) (B) (C) (D) (E)
63. (A) (B) (C) (D) (E)
64. (A) (B) (C) (D) (E)
65. (A) (B) (C) (D) (E)
66. (A) (B) (C) (D) (E)
67. (A) (B) (C) (D) (E)
68. (A) (B) (C) (D) (E)
69. (A) (B) (C) (D) (E)
70. (A) (B) (C) (D) (E)
71. (A) (B) (C) (D) (E)
72. (A) (B) (C) (D) (E)
73. (A) (B) (C) (D) (E)
74. (A) (B) (C) (D) (E)
75. (A) (B) (C) (D) (E)
76. (A) (B) (C) (D) (E)
77. (A) (B) (C) (D) (E)
78. (A) (B) (C) (D) (E)
79. (A) (B) (C) (D) (E)
80. (A) (B) (C) (D) (E)
81. (A) (B) (C) (D) (E)
82. (A) (B) (C) (D) (E)
83. (A) (B) (C) (D) (E)
84. (A) (B) (C) (D) (E)
85. (A) (B) (C) (D) (E)
86. (A) (B) (C) (D) (E)
87. (A) (B) (C) (D) (E)
88. (A) (B) (C) (D) (E)
89. (A) (B) (C) (D) (E)
90. (A) (B) (C) (D) (E)
91. (A) (B) (C) (D) (E)
92. (A) (B) (C) (D) (E)
93. (A) (B) (C) (D) (E)
94. (A) (B) (C) (D) (E)
95. (A) (B) (C) (D) (E)
96. (A) (B) (C) (D) (E)
97. (A) (B) (C) (D) (E)
98. (A) (B) (C) (D) (E)
99. (A) (B) (C) (D) (E)
100. (A) (B) (C) (D) (E)

BIOLOGY EXAM IV

PART A

DIRECTIONS: For each of the following questions or incomplete sentences, there are five choices. Choose the answer which is most correct. Darken the corresponding space on your answer sheet.

1. Select the monosaccharide from the following list of carbohydrates:

   (A) glucose
   (B) lactose
   (C) maltose
   (D) starch
   (E) sucrose

2. The pH of a trout pond contaminated with acid rain is 5. Its hydrogen ion concentration (moles per liter) is:

   (A) .01
   (B) .0001
   (C) .00001
   (D) .0000001
   (E) 1

3. Somatic buffer systems:

(A) are biodegradable

(B) control extracellular pH

(C) make acids

(D) produce nutrients

(E) remove water

4. Two nutrient subunits are bonded together and release a water molecule in which process?

(A) decondensation buildup

(B) dehydration synthesis

(C) dehydrolysis

(D) hydration

(E) hydrolysis

5. Saturated fatty acids:

(A) are the structural building blocks of carbohydrates

(B) have many hydrogens on carbon chains

(C) compose sugars and starches

(D) generally store less energy than protein does

(E) store small amounts of energy

6. Eukaryotic cells are unique cells because they:

(A) are similar to prokaryotic cells

(B) conduct protein synthesis

(C) have a single DNA molecule

(D) have lysosomes

(E) lack ribosomes

Under certain conditions of stability, allelic frequencies and genotypic frequencies remain constant for generations in a sexually-reproducing population.

7. The founders of this principle for genetic equilibrium were:

(A) Boyle and Dalton

(B) Darwin and Wallace

(C) Hardy and Weinberg

(D) Lamarck and Darwin

(E) Watson and Crick

8. A plant cell will not burst if it is in a hypotonic environment due to the presence of its:

(A) cell membrane

(B) cell wall

(C) cytoplasm

(D) nucleus

(E) ribosomes

9. Release of simple substances into the ecosystem for use by producers depends on activity of the:

(A) carnivores
(B) decomposers
(C) herbivores
(D) primary consumers
(E) secondary consumers

10. The particle likely to move most rapidly by diffusion is

(A) hemoglobin
(B) insulin
(C) oxygen
(D) sucrose
(E) starch

11. Levels of blood oxygen are relatively high in a human's:

(A) pulmonary artery
(B) pulmonary vein
(C) right atrium
(D) right ventricle
(E) systemic vein

12. A cell's enzyme system is subjected to a temperature of seventy degrees Celsius for thirty minutes. Its enzymes will experience:

(A) active site loss
(B) allosterism
(C) competitive inhibition
(D) denaturation
(E) modulation

13. The movement of embryonic body cells to produce body shape and form is:

(A) differentiation
(B) diploidy
(C) fertilization
(D) gametogenesis
(E) morphogenesis

14. Cloning is a reproductive process which involves a:

(A) sexual reproductive process
(B) formation of genetic duplicates
(C) combination of meiosis and mitosis
(D) production of genetic variability
(E) yielding of sex cells

15. As energy is passed through a food web:

(A) decomposers cannot break down organisms' wastes
(B) herbivores dominate the last links of the web by numbers of organisms
(C) less useful energy remains available at each successive feeding level
(D) producers feed on remaining nutrients unused by other trophic levels
(E) species numbers increase from link to link in the web

16. An organism's haploid chromosome number is 28. Its diploid chromosome number is:

(A) 7
(B) 14
(C) 28
(D) 56
(E) 84

17. In an alga-minnow-bass-bear food chain, organisms are listed in successive trophic levels. The alga is a producer followed by successive consumer levels. The bass can behave as a:

(A) decomposer
(B) fourth level consumer
(C) producer
(D) secondary consumer
(E) tertiary consumer

18. An organism has body cells normally with 88 chromosomes. With a trisomy at one of its chromosome pairs, its body cell chromosome number is changed to:

(A) 3
(B) 43
(C) 45
(D) 89
(E) 90

19. The complementary half of a DNA double helix strand with base sequence TCGATC is:

(A) ACGATC
(B) AGCTAG
(C) GATCGA
(D) GGCUTA
(E) UGCUAG

Questions 20 - 22 refer to the drawing below.

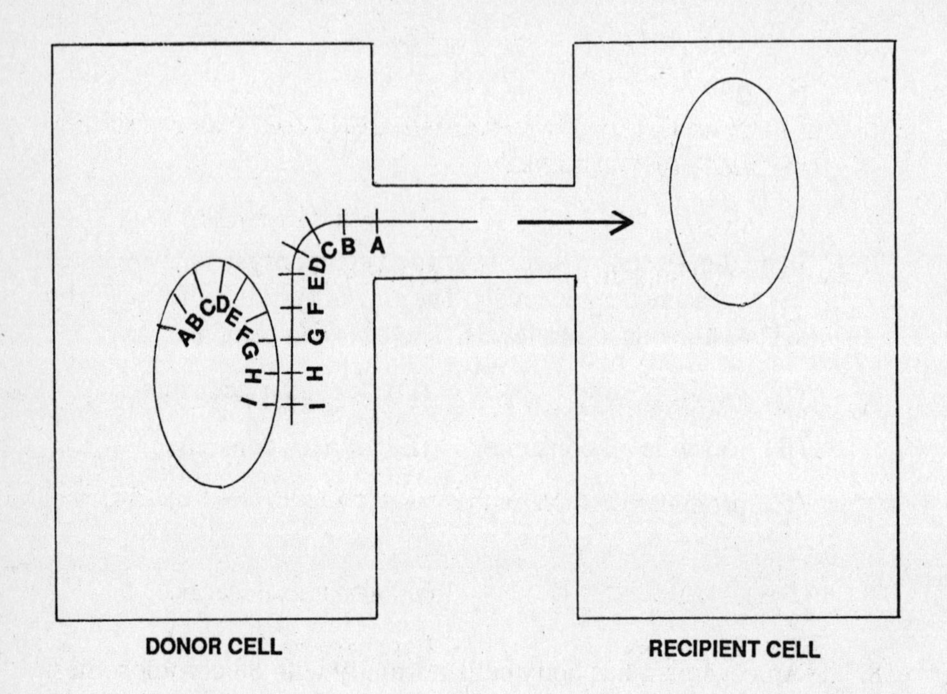

20. Over the next twenty-four hours, which donor gene will the recipient cells receive least out of all the transferred genes?

    (A) gene A
    (B) gene C
    (C) gene D
    (D) gene G
    (E) gene I

21. The donor cell:

    (A) divides
    (B) duplicates genetic material
    (C) has many linear chromsomes

(D) loses its original chromosome

(E) reproduces before transferring chromosomal material

22. If these conjugating bacterial cells are *E. coli*, they could have originated from a human's:

(A) blood
(B) large intestine
(C) skin
(D) small intestine
(E) throat

23. During genetic translation, tRNA derives its name by transferring:

(A) amino acids
(B) fatty acids
(C) fats
(D) monosaccharides
(E) proteins

24. A protein currently synthesized from bacteria through the research of genetic engineering and gene splicing is:

(A) ATP
(B) glucose
(C) growth hormone
(D) hemoglobin
(E) thyroid hormone

25. A tRNA anticodon, UAG, is complementary to an mRNA codon of:

(A) AAG
(B) AUC
(C) AUG
(D) UAG
(E) UUC

Questions 26 - 28 refer to the drawing below.

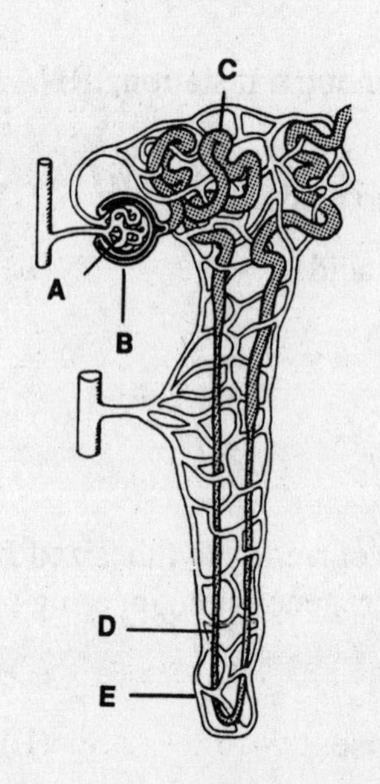

26. Circulatory components of the nephron occur at:

(A) A to B
(B) A to E
(C) B to C
(D) B to D
(E) D to E

27. Components A and B conduct the renal process of:

(A) absorption
(B) filtration
(C) osmosis
(D) reabsorption
(E) secretion

28. In the process of reabsorption, substances move from:

(A) A to B
(B) B to A
(C) B to C
(D) D to E
(E) E to D

29. In human cancer, affected cells tend to:

(A) divide too rapidly
(B) divide too slowly
(C) produce sex cells
(D) remain in a given area
(E) use much more oxygen

30. Messenger RNA:

(A) makes DNA by translation
(B) moves a ribosome along tRNA
(C) orders amino acids while building a protein
(D) orders nucleotides while building a nucleic acid
(E) synthesizes DNA by translation

31. A heterozygous, blood type A person mates with a blood type O individual. A summary of their probabilities for offspring is:

(A) 100% A
(B) 50% A, 50% B
(C) 50% A, 50% O
(D) 50% B, 50% O
(E) 100% O

32. One thousand offspring are counted from a genetic cross. All have the recessive phenotype. The most likely genotypes of the parents are:

(A) AA, AA
(B) AA, Aa
(C) Aa, Aa
(D) Aa, aa
(E) aa, aa

33. Sex-linked traits are:

(A) controlled by genes on the X-chromosome
(B) inherited by sons from their fathers
(C) more common in females
(D) produced by gene pairs in males
(E) usually caused by dominant genes

34. Which is the homozygous recessive condition in humans?

(A) blood type A
(B) brown hair
(C) dark eyes
(D) dwarfism
(E) high cholesterol in the blood

35. Assuming independent assortment and random recombination, an AaBb x aabb mating reproduces all offspring genotypes EXCEPT:

(A) AABB
(B) AaBb
(C) Aabb
(D) aaBb
(E) aabb

36. Select the protozoan class with immobile adults.

(A) Ciliata
(B) Flagellata
(C) Sarcodina
(D) Ctenophora
(E) Sporozoa

37. A protozoan-induced disease is:

(A) acquired immune deficiency syndrome
(B) African sleeping sickness
(C) common cold
(D) polio
(E) spinal meningitis

38. The main reproductive organ in angiosperms is the:

(A) flower
(B) leaf
(C) stem
(D) root
(E) xylem

39. Which is an angiosperm?

(A) fern
(B) mushroom
(C) oak tree
(D) pine tree
(E) spruce

40. Mollusks:

(A) all have a mantle
(B) all have shells
(C) are the spiny-skinned animals
(D) have some evolutionary links to animals
(E) have a closed circulatory system

41. A leg bone is the:

(A) femur
(B) humerus
(C) radius
(D) scapula
(E) ulna

42. Many body internal organs are controlled by the brain's:

(A) cerebellum
(B) hypothalamus
(C) medulla
(D) pons
(E) thalamus

43. Light passes through structures of the eye in the order:

(A) aqueous humor-cornea-lens-vitreous body-retina

(B) aqueous humor-lens-cornea-vitreous body-retina

(C) cornea-aqueous humor-lens-vitreous body-retina

(D) lens-retina-cornea-aqueous humor-vitreous body

(E) retina-aqueous humor-cornea-lens-vitreous body

44. Select the hormone secreted by the posterior lobe of the pituitary gland.

(A) calcitonin
(B) ADH
(C) cortisol
(D) growth hormone
(E) PTH

45. The triceps brachii is an example of a(n):

(A) brain region
(B) digestive chamber
(C) heart chamber
(D) kidney region
(E) skeletal muscle

46. Select the circulatory structure with a low concentration of carbon dioxide in the blood flowing through it.

(A) inferior vena cava
(B) pulmonary artery
(C) pulmonary vein
(D) right ventricle
(E) superior vena cava

47. Exhaled air flows through the tract passages by the sequence:

(A) alveolus-bronchus-trachea-bronchiole-larynx

(B) alveolus-bronchiole-bronchus-trachea-larynx

(C) bronchiole-alveolus-trachea-larynx-bronchus

(D) bronchiole-trachea-larynx-bronchus-alveolus

(E) larynx-trachea-bronchus-bronchiole-alveolus

48. Over twenty-four hours the kidneys eliminate 900 ml. of urine. 500 ml. of liquid is removed by solid wastes and 300 ml. by perspiration and exhalation. Food consumption adds 400 ml. and 800 ml. of liquids are consumed. For water balance, body metabolism contributes:

(A) 100 ml.

(B) 300 ml.

(C) 500 ml.

(D) 700 ml.

(E) 900 ml.

49. Sperm cells are stored and mature in the:

(A) Cowper's gland

(B) epididymis

(C) prostate gland

(D) testis

(E) vas deferens

50. Ovulation in the menstrual cycle corresponds with a peak in the hormone:

(A) estrogen

(B) FSH

(C) ICSH

(D) LH

(E) progesterone

51. Each of the following is a key component of Darwin's principles of natural selection EXCEPT:

(A) competition for limiting resources develops

(B) less population members are produced than can survive

(C) populations' reproduction potential is large

(D) reproductive fitness is the key to a species future

(E) variation is a fact of population makeup

52. Members of the phylum Arthropoda:

(A) contain a backbone

(B) have jointed appendages

(C) include a mantle

(D) lack an exoskeleton

(E) lack appendages

53. The gene for a given trait may survive and increase in time if it is:

(A) adaptive

(B) inferior

(C) heterozygous

(D) homozygous

(E) multiple

54. The most common of the four ABO blood types in the United States is:

(A) A

(B) AB

(C) AO

(D) B

(E) O

55. An autosomal recessive genotype is correctly written as:

(A) A
(B) AA
(C) Aa
(D) a
(E) aa

56. Groups of different population kinds (species) interact to form a:

(A) chemical cycle
(B) community
(C) habitat
(D) niche
(E) organ system

57. The correct ranking of food flow throught the digestive tract is:

(A) esophagus-pharynx-small intestine-stomach-large intestine
(B) pharynx-esophagus-stomach-small intestine-large intestine
(C) large intestine-stomach-small intestine-esophagus-pharynx
(D) stomach-pharynx-esophagus-large intestine-small intestine
(E) small intestine-large intestine-pharynx-stomach-esophagus

58. The soil particle with greatest diameter is:

(A) calcium
(B) clay
(C) humus
(D) sand
(E) silt

59. The most common element in protoplasm by weight is usually:

(A) C
(B) H
(C) O
(D) N
(E) P

60. The last portion of the urinary tract to pass urine is the:

(A) bladder
(B) calyx
(C) pelvis
(D) ureter
(E) urethra

PART B

DIRECTIONS: Each of the following sets of questions refers to a list of lettered choices immediately preceding the question set. Choose the best answer from the list, then darken the corresponding space on your answer sheet. Choices may be used once, more than once, or not at all.

Questions 61 - 64

(A) active transport
(B) exocytosis
(C) osmosis
(D) phagocytosis
(E) pinocytosis

61. cell vacuole expulsion

62. water diffusion through a membrane

63. molecules move from lower to higher concentration

64. "cellular eating"

Questions 65 - 68

(A) deciduous forest

(B) grassland

(C) taiga

(D) tropical rain forest

(E) tundra

65. biome where subsoil is permently frozen, few trees

66. dominated by conifers, cold winters, with a summer warm enough to melt subsoil

67. most abundant rainfall, enormous species diversity

68. relatively low annual rainfall, common in western United States

Questions 69 - 72

(A) conditioning

(B) habituation

(C) imprinting

(D) insight

(E) tropism

69. turning response to a stimulus such as light or gravity, particularly by different growth patterns in palnts.

70. the association, through reinforcement, of a response to a stimulus with which it was not previously associated

71. the strong attachment of newly-born animals with the first moving object they observe

72. a simple learning type with a gradual decline in response to a stimulus through repeated exposure

Questions 73 - 76

(A) operator gene

(B) promotor gene

(C) regulator gene

(D) repressor substance

(E) structural gene

73. makes an RNA message to direct enzyme synthesis directed by cistrons

74. makes enzymes that control metabolism such as lactose in the cell

75. switches on cistron activity

76. synthesizes a molecule that blocks a gene adjacent to structural genes.

PART C

DIRECTIONS: The following questions refer to the accompanying diagram. Certain parts of the diagram have been labelled with numbers. Each question has five choices. Choose the best answer and darken the corresponding space on your answer sheet.

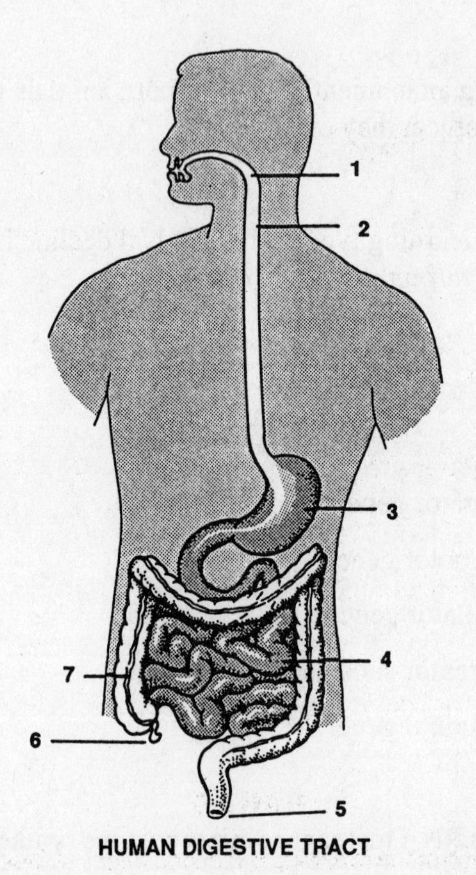

HUMAN DIGESTIVE TRACT

77. A region of this tract shared with the respiratory tract is found at:

(A) 1
(B) 2
(C) 4
(D) 6
(E) 7

78. Absorption of digested nutrients takes place at:

(A) 1
(B) 2
(C) 3
(D) 4
(E) 6

79. Reabsorption occurs at:

(A) 1
(B) 2
(C) 4
(D) 5
(E) 7

80. The esophagus is:

(A) 1
(B) 2
(C) 5
(D) 6
(E) 7

PART D

DIRECTIONS: Each of the following questions refer to a laboratory or experimental situation. Read the description of each situation, then select the best answer to each question. Darken the corresponding space on the answer sheet.

Question 81 - 84 refer to the dichotomous key and drawings below. Take each organism through the key for correct genus identification.

1. a. long, slender body form - go to step 2

   b. short, thick body form - go to step 3

2. a. body has stripes - *Albus*

   b. body lacks stripes - *Taxus*

3. a. body has stripes - *Leus*

   b. body lacks stripes - *Unus*

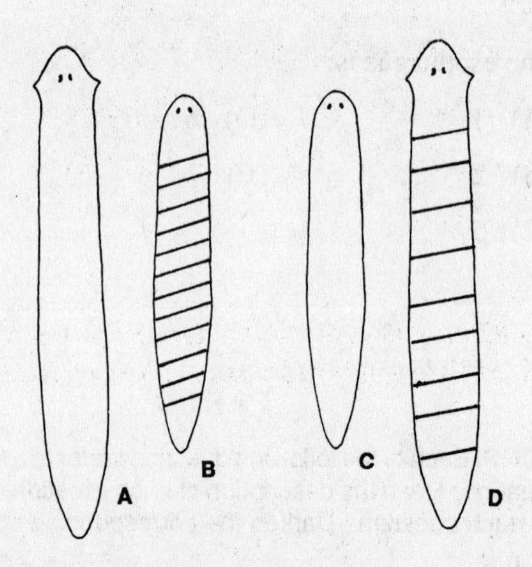

81. Organism A belongs to genus:

(A) *Albus*

(B) *Leus*

(C) *Taxus*

(D) *Xenus*

(E) *Unus*

82. Organism B belongs to genus:

(A) *Albus*
(B) *Leus*
(C) *Taxus*
(D) *Xenus*
(E) *Unus*

83. Organism C belongs to genus:

(A) *Albus*
(B) *Leus*
(C) *Taxus*
(D) *Xenus*
(E) *Unus*

84. Organism D belongs to genus:

(A) *Albus*
(B) *Leus*
(C) *Taxus*
(D) *Xenus*
(E) *Unus*

Questions 85 and 86 refer to the karyotype below. A karyotype is an organized listing of chromosomes into related groups.

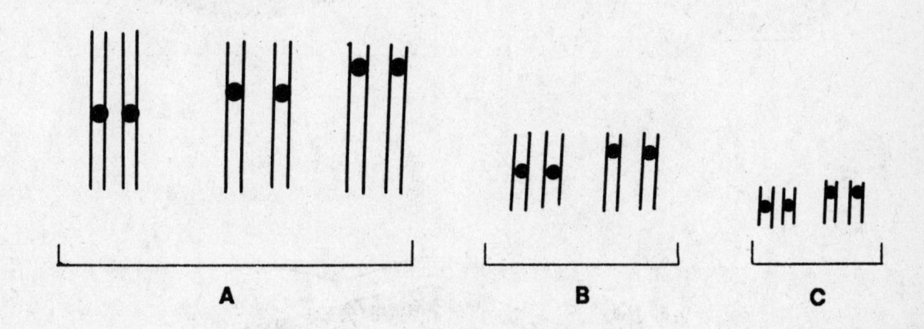

Karyotype

85. The diploid chromosome number for this karyotyped species is:

(A) 2
(B) 4
(C) 7
(D) 14
(E) 28

86. The haploid chromosome number for this karyotyped species is:

(A) 1
(B) 2
(C) 7
(D) 14
(E) 21

Questions 87 and 88 refer to the drawings of the following microscopic protozoans.

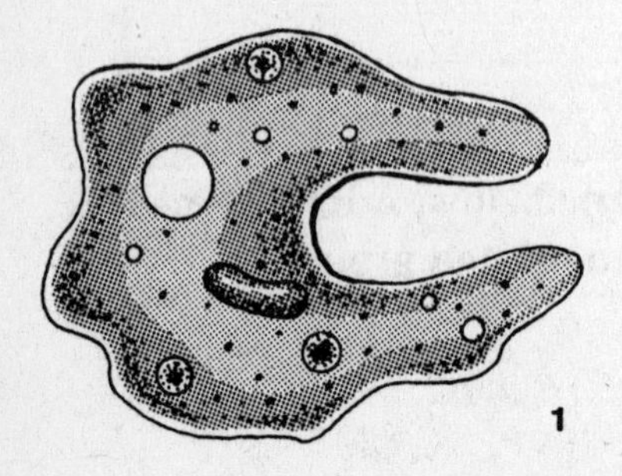

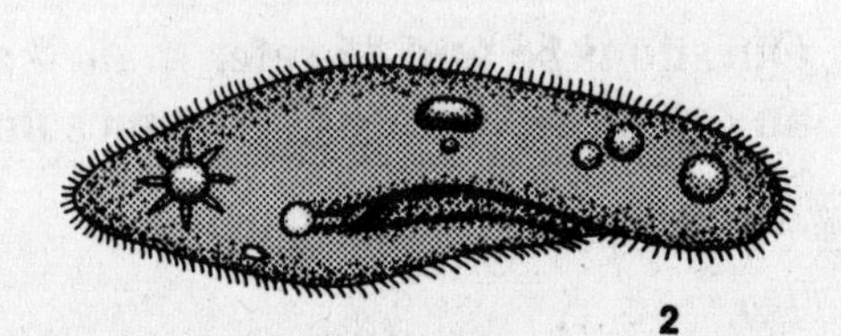

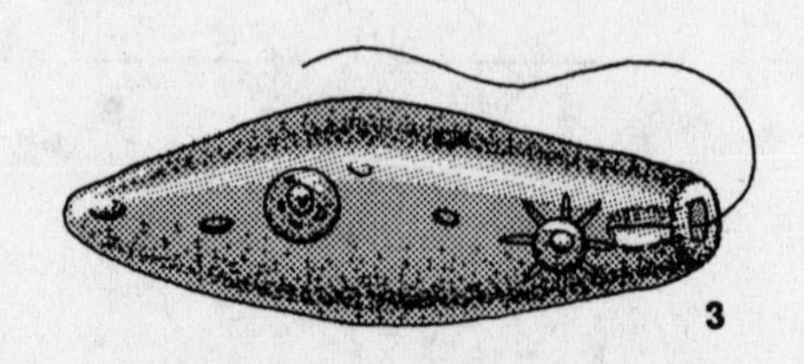

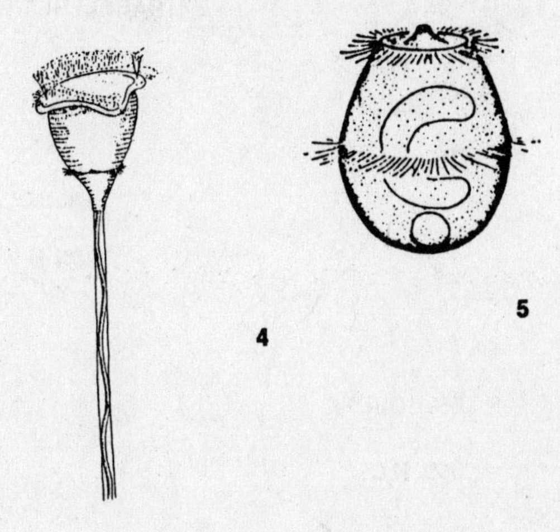

87. The *Euglena* is organism number:

(A) 1
(B) 2
(C) 3
(D) 4
(E) 5

88. The *Paramecium* is organism number:

(A) 1
(B) 2
(C) 3
(D) 4
(E) 5

Questions 89 and 90 refer to the cell below.

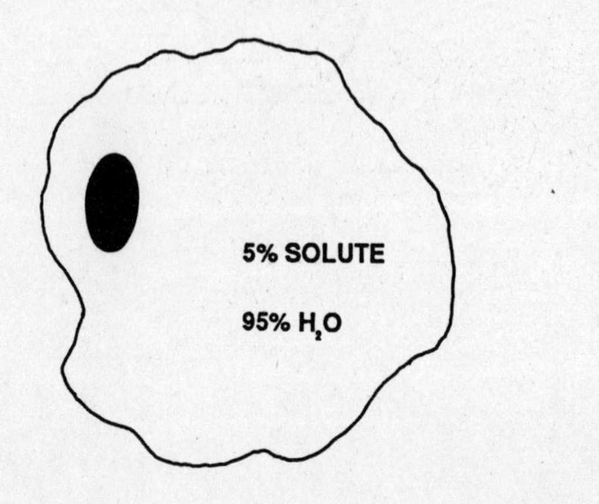

89. Select the correct statement about the cell and its setting.

    (A) The cell will gain solute.

    (B) The cell will lose water.

    (B) The extracellular environment is hypotonic.

    (D) The intracellular environment is hypertonic.

    (E) The two environments are isotonic.

90. Select the concentration hypotonic to both the extracellular and intracellular setting.

    (A) 3%

    (B) 5%

    (C) 7%

    (D) 8%

    (E) 10%

Questions 91 - 94 refer to the microscopic section of the plant root.

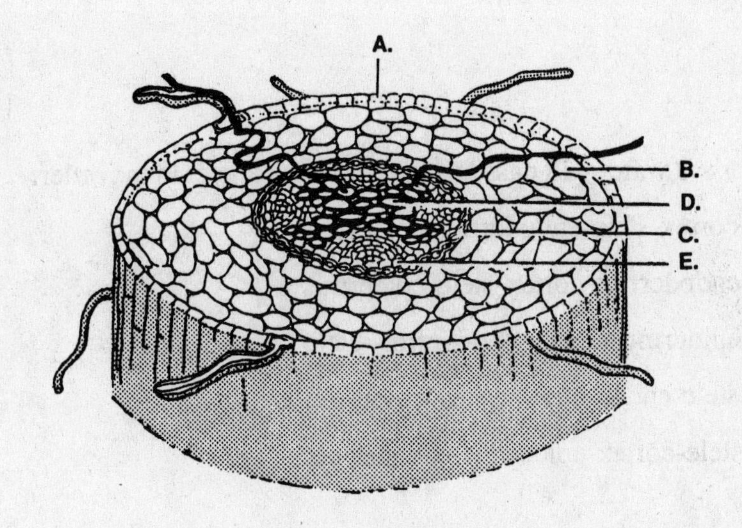

91. Water and minerals are conducted by structures:

(A) A
(B) B
(C) C
(D) D
(E) E

92. A major function of B is:

(A) photosynthesis
(B) repel water
(C) store starch
(D) transport organic solute
(E) transport water

93. The endodermis is at:

(A) A
(B) B
(C) C
(D) D
(E) E

94. Absorbed water will pass through this root's layer in the order:

(A) cortex-stele-endodermis-epidermis
(B) endodermis-cortex-stele-epidermis
(C) epidermis-cortex-endodermis-stele
(D) stele-endodermis-cortex-epidermis
(E) stele-cortex-epidermis-endodermis

Questions 95 and 96 refer to the graphs below of the menstrual cycle of a guinea pig.

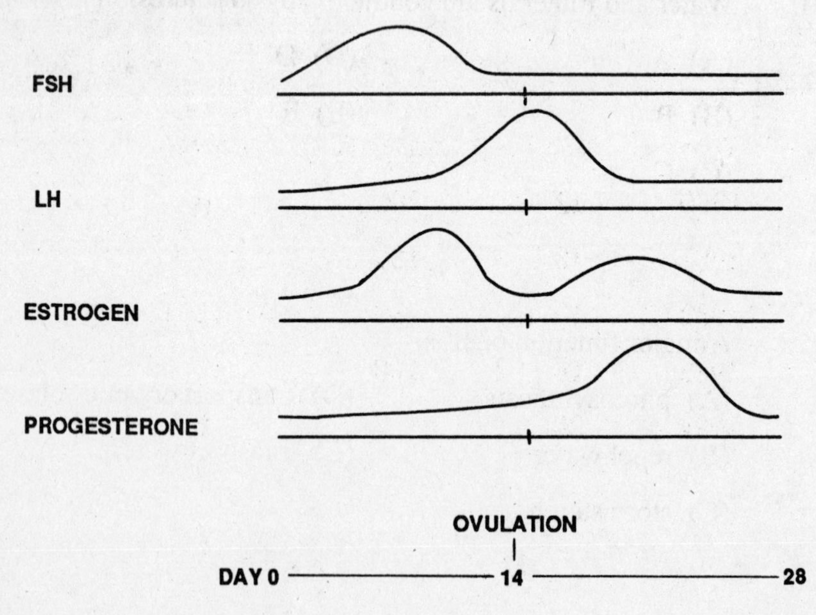

95. The hormone stimulating ovulation is:

(A) estrogen
(B) FSH
(C) ICSH
(D) LH
(E) progesterone

96. Progesterone's possible effect is to:

(A) dominate the first half of the menstrual cycle
(B) keep this particular cycle running indefinitely
(C) prepare the uterus for possible embryo acceptance
(D) stimulate increased production of FSH
(E) work against the effect of estrogen

Questions 97 - 100 refer to the data below computed from a field study.

Year	Acres	(No. of) Pheasants	(No. of) Foxes
1980	12	200	6
1981	18	180	14
1982	10	80	18
1983	16	100	12
1984	20	140	8

97. A possible symbiotic process relating population change in the species displayed is:

(A) amensalism
(B) commensalism
(C) mutualism
(D) parasitism
(E) predation

98. The greatest change in population density from either species occurs by the year:

(A) 1980
(B) 1981
(C) 1982
(D) 1983
(E) 1984

99. The population density of pheasants decreases from:

(A) 1980-1984
(B) 1981-1984
(C) 1982-1983 only
(D) 1980-1983 only
(E) the density never decreases

100. What would you expect to happen to the population density of foxes if the pheasants were eating a substance toxic to foxes, but not to themselves throughout this study (1980-1984)?

(A) The fox population density would remain constant
(B) The fox population density would follow the same pattern as observed
(C) The fox population density would decrease sharply
(D) The fox population density would increase
(E) The foxes would not eat affected pheasants

BIOLOGY
TEST IV

ANSWER KEY

1.	A	34.	D	67.	D
2.	C	35.	A	68.	B
3.	B	36.	E	69.	E
4.	B	37.	B	70.	A
5.	B	38.	A	71.	C
6.	D	39.	C	72.	B
7.	C	40.	A	73.	B
8.	B	41.	A	74.	E
9.	B	42.	B	75.	A
10.	C	43.	C	76.	C
11.	B	44.	B	77.	A
12.	D	45.	E	78.	D
13.	E	46.	C	79.	E
14.	B	47.	B	80.	B
15.	C	48.	C	81.	C
16.	D	49.	B	82.	B
17.	D	50.	D	83.	E
18.	D	51.	B	84.	A
19.	B	52.	B	85.	D
20.	E	53.	A	86.	C
21.	B	54.	E	87.	C
22.	B	55.	E	88.	B
23.	A	56.	B	89.	B
24.	C	57.	B	90.	A
25.	B	58.	D	91.	D
26.	B	59.	C	92.	C
27.	B	60.	E	93.	C
28.	D	61.	B	94.	C
29.	A	62.	C	95.	D
30.	C	63.	A	96.	C
31.	C	64.	D	97.	E
32.	E	65.	E	98.	B
33.	A	66.	C	99.	D
				100.	C

BIOLOGY
TEST IV

DETAILED EXPLANATIONS OF ANSWERS

1. (A)

Glucose is a simple sugar, or monosaccharide. Lactose, sucrose, and maltose are dissacharides; they each have two monosaccharides in their molecular structure. Starch is a polysaccharide.

2 (C)

ph= $-\log [H^+]$, in which $[H^+]$ is the concentration of protons (in moles per liter) in the solution. $-\log [.00001] = 5$.

3. (B)

The pH of the human body's extracellular fluid remains remarkably constant (7.4), in spite of environmental conditions that constantly exist to disturb it. For example, acid-containing foods (which donate hydrogen ions) are met by buffering compounds that counteract the excess acidity and prevent a movement toward a lower, more acidic, pH.

4. (B)

This choice represents a formation of molecule by the bonding of two smaller molecules (synthesis) accompanied by water liberation (dehydration). The H^+ and OH^- that are removed from the molecules leave a vacant bond on each. The two molecules are then combined in order to form a dimer.

5. (B)

The term "saturated" refers to saturation of the fatty acid's carbon chain with hydrogen. These hydrogens are a source of energy when removed by oxidation during metabolism. Fatty acids, building blocks of fats, contain more hydrogen per gram than the other listed molecule types, such as carbohydrates and proteins.

6. (D)

Eukaryotic cells are quite different from prokaryotic cells; they have many unique characteristics that allow one to distinguish them from prokaryotic cells. All eukaryotic cells have well-defined nuclei bound by nuclear membranes. Prokaryotic cells do not. Eukaryotic nuclei contain genetic information in the form of chromosomes, large molecules of DNA combined with associated histone proteins. Prokaryotic cells have single circular molecules of DNA. Eukaryotic cells also have specialized organelles, such as mitochondria, chloroplasts, Golgi complexes, lysosomes, and endoplasmic reticula. Prokaryotic cells do not. They do, however, possess ribosomes (which are smaller than the ribosomes of eukaryotes), which allow them to synthesize proteins that they need.

7. (C)

This is a historical fact. They provided a set of conditions that would maintain constant gene frequencies in a population. If natural selection does not affect a large, isolated population, and if no mutations occur within the population, then no change in gene frequencies will occur; the population will not evolve.

8. (B)

The plant cell is hypertonic, and so water will enter the cell. This is because water is more concentrated outside of the cell than it is inside the cell. Water molecules will travel from a region of greater concentration to one of lower concentration. However, if too much water were to enter the plant cell, then the cell would rupture. A cell wall prevents the plant from doing so by constraining the cell.

9. (B)

Decomposers are the bacteria, fungi, and other microorganisms that digest organisms' carcasses and waste. Their actions return elements from an ecosystem's biotic component back to the abiotic sector of the environment.

10. (C)

Light, gaseous particles move most rapidly. Larger, more massive macromolecules are far less likely to do so. Oxygen is the lightest molecule listed. All other choices represent molecules of much greater molecular weights.

11. (B)

Pulmonary veins bring blood back to the human heart's left atrium from the lungs, rejuvenating the blood with oxygen and removing some of the gaseous waste product, carbon dioxide. The lungs oxygenate blood and remove carbon dioxide. Blood drained from the body by systemic veins enters the heart's right chambers, and continues on through the pulmonary artery, where it is located prior to this reversal carried out by the lungs.

12. (D)

Denaturation is the physical breakdown of a protein under extremes of temperature or pH. Allosterism and modulation are normal changes in the enzyme's shape to accommodate its fitting to a substrate. The active site is the part of an enzyme binding to a substrate. The active site is the part of the enzyme binding to a substrate. Competitive inhibition refers to inhibition whereby a molecule resembling the substrate enters and binds with the active site. This decreases the availability of the active site for the normal substrate and it decreases the reaction rate.

13. (E)

This is a fact of studying development. Differentiation is cell specialization. Gametogenesis is sex cell production. Fertilization is sex cell union. Diploidy refers to chromosomes in developing cells occurring in pairs. The genes contained on the chromosomes carry instructions to direct development.

14. (B)

This is a simple description of cloning. If body cells are undifferentiated (such as the zygote), they can produce new organisms by cell division, asexually. Cloning has been done commonly with plant structures and cells from the bellies of salamanders, for example.

15. (C)

Producers and the herbivores that feed on them start a food web. Each step of food chain energy conversion leads to dissipation of some energy into useless heat. This limits numbers that can be supported at each successive link, as less energy to support them becomes available. Decomposers act on organisms and their wastes throughout.

16. (D)

Sex cells are haploid (half number). The diploid number in somatic cells is double the haploid number in a species. In somatic cells, two copies of each chromosome type are present.

17. (D)

The bass is a second-level feeder. It is not at a third step nor fourth step of feeding. It is also not a producer starting the chain, as is the alga.

The minnow is an herbivore, primary consumer. The bear is a third-order consumer.

18. (D)

Trisomy indicates that there is one extra chromosome in a pair of chromosomes. Thus, the pair is actually a trio (3 chromosomes). If the normal number of chromosomes is 88 and there is an extra chromosome, there are 89 chromosomes in the cells of the individual.

19. (B)

Memorization of the base-pairing rules of A-T and G-C is necessary to answer this. This six base DNA readout has only one possible complement. Thus the base T calls A in the other half. C attracts G, etc.

20. (E)

More and more chromosomal material will enter the recipient cell as time accrues.

21. (B)

The donor cell first duplicates the gene sequence of its original, single circular chromosome. This newly-made chromosome is transferred across the cytoplasmic bridge.

22. (B)

E. coli is a normal floral inhabitant of the human colon, or large intestine. It derives food and a habitat without harming the human body. It is a frequently-used microorganism in genetic research because of its advantages in study: short generation time, large number of offspring, haploid genetic structure.

23. (A)

Amino acids have no direct affinity for messenger RNA at the ribosome. For assemblage into a protein, they must first couple to a tRNA molecule that is compatible to the particular part of the message and that carries them to it. The tRNA is compatible by matching its complementary anti-codon to the correct mRNA codon directing translation at the ribosome.

24. (C)

Growth hormone, insulin, and interferon are all current products now synthesized bacterially by gene transplants.

25. (B)

The base-pairing RNA rules of A-U and C-G must be memorized to answer this question.

26. (B)

The nephron is the microscopic kidney unit. It has two capillary networks: the glomerulus (A) inside the Bowman's capsule and the peritubular capillaries (E) surrounding the nephron's tubular portion.

27. (B)

Filtration is the bulk flow, unselective passage of blood plasma substances from the glomerulus into the Bowman's capsule. This is the first step of renal physiology and introduces blood substances into the urinary tract at the microscopic level. Reabsorption involves reclaiming blood substances from the filtrate back into the blood. Secretion is the final renal step eliminating substances from blood into the nephron tract. The other choices are not renal processes.

28. (D)

After filtration, substances are monitored by kidney nephrons and returned to the blood at high percentage rates as needed. This reabsorption process occurs from nephron tubules back into the blood in the peritubular capillaries.

29. (A)

Cancer is mitosis of an organism's body cells at abnormally high rates. The body cells, such as white blood cells in leukemia, divide more rapidly than needed to merely replace worn-out cells. These rapidly-dividing cells tend to break away and wander, seeding other body regions. Their continual relocation and rapid division spreads the cancer.

30. (C)

mRNA bears a complementary message reflecting the base readout of an activated gene's (DNA) content. The message orders amino acids, one-by-one in their sequence specified by a gene.

31. (C)

A heterozygous blood type A (I^Ai) person will offer a dominant A or recessive O gene through the sex cell with equal probability. Either combines with the other parent's O gene, yielding a type A or type O offspring.

	I^A	i
i	I^Ai	ii
i	I^Ai	ii

32. (E)

Homozygous recessive parents can donate only recessive alleles in sex cells and thus reproduce homozygous recessive offspring. All other choices produce 50, 75, or 100% dominant-appearing offspring. The large number of counted offspring statistically affirms the inheritance of only recessive genes.

33. (A)

Sex-linked traits, such as hemophilia and color blindness, are sometimes called X linked because they are controlled by recessive genes on X chromosomes. Sons inherit this gene from their mother who is usually a carrier for the trait i.e. heterozygous. The expression of such traits are more common in males. They have only a single gene since they possess only one X chromosome. The Y chromosome does not have a second gene. Therefore, a recessive allele on the X chromosome is not masked by a second gene.

34. (D)

aa x aa can yield only aa offspring (dwarfs).

35. (A)

Study a Punnett square. The first parent can offer an AB gamete but the second parent cannot with its genotype, for an AABB recombination.

AaBb x aabb

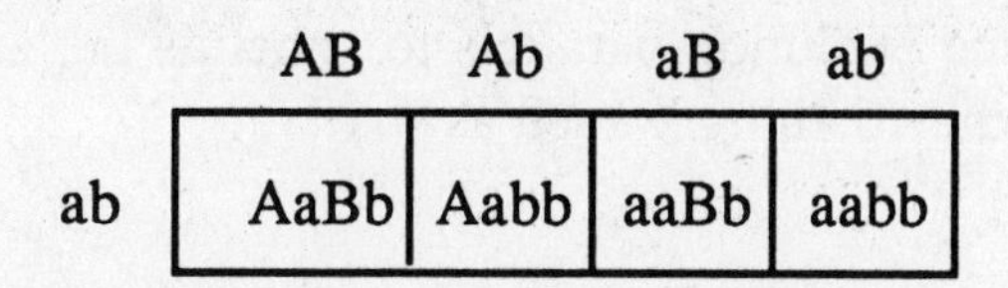

	AB	Ab	aB	ab
ab	AaBb	Aabb	aaBb	aabb

36. (E)

Ciliata members move by cilia. Members of Flagellata employ flagella and Sarcodina members use amoeboid motion. Ctenophora organisms are not protozoans. They are comb-jellies, closely allied to the jellyfish group. An immobile sporozoan, *Plasmodium*, invades the human bloodstream to cause malaria.

37. (B)

The other four diseases are viral-induced.

38. (A)

The leaf is photosynthetic. The stem supports the plant and contains xylem, which is a transport mechanism. The root absorbs H_2O and nutrients, transports materials, and anchors the plant. The flower produces sex cells and offers a fertilization site.

39. (C)

Ferns are lower vascular plants. They bear fronds which produce spores. Mushrooms belong to the Kingdom Fungi. The pine tree and spruce are both gymnosperms. They bear cones as reproductive structures. The oak tree is an angiosperm. It bears flowers and "hidden" seeds.

40. (A)

The mantle is a membrane for respiration and cell secretion. Advanced members, octopus and squid, have lost their shell but retain the mantle. They have some evolution parallels to annelids but are not echinoderms - spiny skinned animals such as starfish.

41. (A)

The humerus (upper arm) and the radius (lateral forearm) and ulna (medial forearm) are upper appendage bones. The scapula, shoulder blade, articulates the humerus to the axial skeleton.

42. (B)

This lower forebrain function is multifaceted, controlling body temperature, blood sugar level and extracellular fluid levels. The medulla is at the base of the brainstem with the pons continuing above it. The cerebellum is posterior to both. All are hindbrain structures. The thalamus is above the hypothalamus in the forebrain.

43. (C)

The outward bulging, clear cornea intercepts light rays which then pass through the water-like aqueous humor in the cavity behind it. The football-shaped lens next refracts light rays, concentrating them on the retina's focal point as they pass through the jellylike vitreous body.

44. (B)

This hormone (antidiuretic) facilitates water reabsorption by the kidney. Calcitonin (thyroid) and PTH (parathyroid hormone) control blood calcium. Cortisol comes from the adrenal cortex with many effects (i.e. - raise blood sugar). The growth hormone is produced from the pituitary's anterior lobe.

45. (E)

The triceps is the major muscle on the back of the upper arm. It pulls the forearm bones to straighten them out from the elbow joint-extension. It opposes in motion the biceps brachii muscle on the front of the upper arm. It produces forearm bending or flexion.

46. (C)

This vessel brings rejuvenated blood (higher oxygen and lower carbon dioxide) back to the heart's left side. The superior (head, neck, arms) and inferior (the balance) vena cavae bring blood lower in oxygen and higher in carbon dioxide through the heart's right side prior to its trip to the lungs.

47. (B)

Exhaled air leaves microscopic alveoli and enters the tubelike bronchioles and larger bronchi as air leaves the lungs. Remaining respiratory structures liberate air in the order given by the choice. Choice E gives the correct sequence for air inhalation.

48. (C)

Total outputs equal 1700 ml. The two given inputs total 1200 ml., leaving 500 ml. produced by metabolism for an input-output balance.

49. (B)

This structure resides somewhat on top of and in back of the testis to store maturing sperm cells after their synthesis in the testis.

50. (D)

LH, the luteinizing hormone from the pituitary gland, peaks 14 days before the menstrual cycle's end to key ovulation. It also converts the follicle, having released the migrating ovum, into a corpus luteum. FSH is another pituitary hormone. Its upswing initiates the cycle. Estrogen and progesterone are ovarian hormones maintaining the uterus. ICSH is a male hormone.

51. (B)

Darwin stated that populations reproduce many more members than can possibly be supported by environmental resources. Thus, competition occurs for limiting resources. Those best equipped to survive have a higher probability to leave offspring and contribute to the gene pool of the next generation.

52. (B)

These invertebrates, with exoskeleton and jointed appendages, include insects as their largest class. A mantle, found in mollusks, is lacking.

53. (A)

Adaptive means to promote survival. Organisms adaptively equipped with their gene have a higher probability to leave offspring and influence the future gene pool.

54. (E)

The approximate frequency is O (45%),A(40%), B (10%), and AB (4-5%).

55. (E)

A recessive allele is always indicated by a lower case letter (a, for example). A dominant allele is always indicated by a capital letter (A, for example). A genotype indicates the genetic composition. The autosomal traits are in pairs in the normal cell. (One is found on each of the 2 chromosomes of a given chromosome pair in the diploid individual). The genotype of the recessive individual is therefore aa. AA is the genotype for the homozygous dominant individual. Recessives always contain two recessive alleles for a given trait.

56. (B)

This is a strict definition of an ecosystem's community.

57. (B)

The pharynx is the throat. Swallowed food moves through the esophagus (a ten- inch muscular tube). The esophagus introduces food into the stomach. The small intestine is about twenty feet long and is attached to the stomach. The next organ is the large intestine, (five to six feet in length).

58. (D)

Silt is smaller and clay is smallest. Calcium is an element and humus is a collective term for decaying organic matter in soil.

59. (C)

O stands for oxygen and usually constitutes some sixty percent worth of living material or protoplasm among the four most common elements :O, C (carbon) (18%), N (nitrogen) (4%), and H, (hydrogen) (10%).

60. (E)

The sequence is kidney - ureter - bladder- urethra. The kidney forms the urine, sending it down the ureter (a ten-inch tube) into the bladder for storage. The urethra is a tract draining the bladder for urine elimination. The calyx is a kidney structure.

61. (B) 62. (C) 63. (A) 64. (D)

Through the process of diffusion, molecules are spread passively. Such molecules move from higher to lower concentration. Energy is not expended. However, they move oppositely with cell energy expenditure during active transport. When water diffuses through a semipermeable membrane, osmosis occurs. Phagocytosis involves a mobile cell, amoeboid or white blood cell for example, surrounding and engulfing a foreign particle. Exocytosis is the escape of materials from a cell. Pinocytosis is the taking in of dissolved molecular food materials by adhering them to the plasma membrane to form vacuoles. It is a form of active transport.

65. (E) 66. (C) 67. (D) 68. (B)

Each correct answer is a definition of a worldwide ecosystem or biome. The far northern latitude of the tundra support few trees while continually cold temperatures keep the subsoil frozen throughout the year. The taiga is south of the tundra in the northern hemisphere and is warmer with summers that offer growing seasons for larger trees. Grasslands, prominent in the Western U.S., are dry and typed by brushlike vegetation. The lush tropical rainforests may receive over 100 inches of rainfall annually. A small plot of land may contain hundreds of species.

69. (E) 70. (A) 71. (C) 72. (B)

Tropisms include the turning responses of plants (gravitropism) and to light (phototropism). Pavlov's experiments are famous for the discovery of conditioning. Conditioned dogs eventually recognized the associated site of food with a sounded bell. The dogs soon salivated to the bell stimulus alone. The behaviorist Konrad Lorenz observed imprinting when newborn ducklings would accept a toy car or human figure as a mother figure, if these were the first objects they saw. Repeated bombardment of a stimulus, particularly an irrelevant one, gradually loses its effect in habituation. Consider the "fade out" of an originally new aroma in a room.

73. (B) 74. (E) 75. (A) 76. (C)

All choices are parts of the operon model of gene activity. A successive group of structural genes (cistrons) encode the successive enzymes that catalyze the step-by-step metabolism of a substrate entering a cell. An example of a substrate is the sugar lactose. An adjacent operator gene of the operon switches the activity of the cistrons off and on as needed for metabolism. A nearby regulator gene makes a repressor molecule that binds to the operator gene, turning off the switch and shutting down substrate metabolism. An inducer, the actual substrate such as lactose, can bind with the repressor molecule and allow the operator gene to switch on the metabolic pathway. Another gene, the promotor, makes an enzyme to build the RNA, transcribed off of the structural gene's DNA, that will direct enzyme synthesis.

77. (A) 78. (D) 79. (E) 80. (B)

The pharynx (throat) of the upper digestive tract also accepts air in the upper respiratory tract at 1. The ten-inch esophogus (2), moves swallowed food into the stomach at 3. In the 20 foot small intestine at 4, chemical digestion is completed and the subunits of larger ingested molecules can now enter the blood by absorption for body needs. The next digestive compartment is the large intestine (7) where mostly water is reclaimed into the blood to prevent body dehydration - (reabsorption). The appendix (6) is a vestigial organ. The anus (5) is the site of elimination of solid waste at the end of the digestive tract.

81. (C)

This snake is long and slender, lacking stripes.

82. (B)

This snake is short and thick, with stripes.

83. (E)
This snake is short and thick, lacking stripes.

84. (A)
This snake is long and slender, with stripes.

85. (D)
Diploid species chromosome numbers are computed by the formula 2n where n equals the number of different kinds of chromosomes. The number of different kinds of chromosomes displayed in the karyotype of this species is seven. Note the distinctness of each chromosome by size and centromere location. Seven times two equals fourteen.

86. (C)
Haploid species chromosome numbers are computed by the formula 1 n. There are seven different types of chromosomes displayed in this karyotype.

87. (C) 88. (B)
Protozoans are distinguishable and classified by their means of motility. Note the flagellum of the protozoan's cell for organism #3. *Euglena* is a flagellated organism. The flagellum whips *Euglena* along. *Paramecium*, organism #2, is ciliated. Cilia are numerous, short hairlike projections that beat to propel the cell. *Amoeba*, organism #1 moves by pseudopodia. *Stentor* (4), is a ciliate (same phylum as *Paramecium*) and attaches itself to rocks via a long stalk. Organism 5 is a dinoflagellate. Certain species of this algal group cause red tides. A red tide is simply a population explosion of these reddish or brown species. This huge population, however, produces dangerous toxins.

89. (B)

The extracellular-intracellular settings are not isotonic (equal in solute concentration). The extracellular environment is hypertonic, higher in solute concentration. The intracellular environment is hypotonic, 5% is less than 8%. Water will flow from the cell, 95%, to the outside, 92%, by osmosis. By osmosis water flows from a higher to lower concentration through a membrane permeable to it. The solute gradient goes into the cell, from 8% to 5%. It will not pass in, however, because the membrane is impermeable to it. Hypertonic and hypotonic are relative terms. For example, a 5% solute solution is hypotonic to a 10% solution but hypertonic to a 2% solution.

90. (A)

Only 3% is less than both 5% and 8%, thus, hypotonic to both solute concentrations.

91. (D)

The larger-bored xylem cells alternate with phloem conducting tissue, resembling the points of a star. Both are in the conducting core or stele of the root. Xylem specializes in water and mineral transport. Phloem functions more for conduction of organic molecules.

92. (C)

Cells at layer B are the large parenchymal cells of the root cortex. They specialize for starch storage, not for transport or photosynthesis.

93. (C)

The endodermis is a single layer of cells defining the conducting core or stele.

94. (C)

Soil water passes from the outside, inward. The root layers in this direction are the outer epidermis, thick cortex, endodermis defining the stele, and the stele itself.

95. (D)

Ovulation is marked on the menstrual cycle graph and corresponds to a peaking of LH (luteinizing hormone) levels. Its peak keys the rupture of an ovarian sex cell and its release into the oviduct. The remaining ovarian follicle, a jacket of cells that had encased the sex cell, is converted into the corupus luteum by the LH signal. FSH (follicle stimulating hormone) initiates the cycle by its increase. Estrogen and progesterone are ovarian hormones maintaining the uterus. ICSH is a male hormone.

96. (C)

Progesterone's levels dominate in the phase of the cycle beyond ovulation. By then the female sex cell has entered the oviduct for possible fertilization. If this occurs, it arrives in the uterus as an early embryo. Progesterone, the "hormone of pregnancy," prepares the uterine lining for this possible acceptance.

97. (E) 98. (B) 99. (D) 100. (C)

The fox is a possible predator of pheasants in an ecosystem food chain. Notice how a decrease in pheasant prey correlates with annual gains in fox success by numbers. The density of a population is determined by dividing the number of individuals in the population by the number of acres. The largest change in the population density of either organism occurs in the year 1981. The pheasant population density decreases from 16 (in 1980) to 10 (in 1981). The pheasant population density decreases from 16 to 10 to 8 to 6 from 1980 to 1983, respectively. The population density of pheasants increases slightly in 1984.

If the pheasants were eating something toxic to the fox but not to themselves, one might expect the fox population density to increase slightly in response to the large number of pheasants. However, since the pheasants are poisonous, the fox population would suddenly decrease. The decrease would be quite rapid. Under such conditions, the number of pheasants would increase rapidly due to the decreased number of predators.

BIOLOGY

TEST V

BIOLOGY
ANSWER SHEET

1. (A) (B) (C) (D) (E)
2. (A) (B) (C) (D) (E)
3. (A) (B) (C) (D) (E)
4. (A) (B) (C) (D) (E)
5. (A) (B) (C) (D) (E)
6. (A) (B) (C) (D) (E)
7. (A) (B) (C) (D) (E)
8. (A) (B) (C) (D) (E)
9. (A) (B) (C) (D) (E)
10. (A) (B) (C) (D) (E)
11. (A) (B) (C) (D) (E)
12. (A) (B) (C) (D) (E)
13. (A) (B) (C) (D) (E)
14. (A) (B) (C) (D) (E)
15. (A) (B) (C) (D) (E)
16. (A) (B) (C) (D) (E)
17. (A) (B) (C) (D) (E)
18. (A) (B) (C) (D) (E)
19. (A) (B) (C) (D) (E)
20. (A) (B) (C) (D) (E)
21. (A) (B) (C) (D) (E)
22. (A) (B) (C) (D) (E)
23. (A) (B) (C) (D) (E)
24. (A) (B) (C) (D) (E)
25. (A) (B) (C) (D) (E)
26. (A) (B) (C) (D) (E)
27. (A) (B) (C) (D) (E)
28. (A) (B) (C) (D) (E)
29. (A) (B) (C) (D) (E)
30. (A) (B) (C) (D) (E)
31. (A) (B) (C) (D) (E)
32. (A) (B) (C) (D) (E)
33. (A) (B) (C) (D) (E)
34. (A) (B) (C) (D) (E)
35. (A) (B) (C) (D) (E)
36. (A) (B) (C) (D) (E)
37. (A) (B) (C) (D) (E)
38. (A) (B) (C) (D) (E)
39. (A) (B) (C) (D) (E)
40. (A) (B) (C) (D) (E)
41. (A) (B) (C) (D) (E)
42. (A) (B) (C) (D) (E)
43. (A) (B) (C) (D) (E)
44. (A) (B) (C) (D) (E)
45. (A) (B) (C) (D) (E)
46. (A) (B) (C) (D) (E)
47. (A) (B) (C) (D) (E)
48. (A) (B) (C) (D) (E)
49. (A) (B) (C) (D) (E)
50. (A) (B) (C) (D) (E)
51. (A) (B) (C) (D) (E)
52. (A) (B) (C) (D) (E)
53. (A) (B) (C) (D) (E)
54. (A) (B) (C) (D) (E)
55. (A) (B) (C) (D) (E)
56. (A) (B) (C) (D) (E)
57. (A) (B) (C) (D) (E)
58. (A) (B) (C) (D) (E)
59. (A) (B) (C) (D) (E)
60. (A) (B) (C) (D) (E)
61. (A) (B) (C) (D) (E)
62. (A) (B) (C) (D) (E)
63. (A) (B) (C) (D) (E)
64. (A) (B) (C) (D) (E)
65. (A) (B) (C) (D) (E)
66. (A) (B) (C) (D) (E)
67. (A) (B) (C) (D) (E)
68. (A) (B) (C) (D) (E)
69. (A) (B) (C) (D) (E)
70. (A) (B) (C) (D) (E)
71. (A) (B) (C) (D) (E)
72. (A) (B) (C) (D) (E)
73. (A) (B) (C) (D) (E)
74. (A) (B) (C) (D) (E)
75. (A) (B) (C) (D) (E)
76. (A) (B) (C) (D) (E)
77. (A) (B) (C) (D) (E)
78. (A) (B) (C) (D) (E)
79. (A) (B) (C) (D) (E)
80. (A) (B) (C) (D) (E)
81. (A) (B) (C) (D) (E)
82. (A) (B) (C) (D) (E)
83. (A) (B) (C) (D) (E)
84. (A) (B) (C) (D) (E)
85. (A) (B) (C) (D) (E)
86. (A) (B) (C) (D) (E)
87. (A) (B) (C) (D) (E)
88. (A) (B) (C) (D) (E)
89. (A) (B) (C) (D) (E)
90. (A) (B) (C) (D) (E)
91. (A) (B) (C) (D) (E)
92. (A) (B) (C) (D) (E)
93. (A) (B) (C) (D) (E)
94. (A) (B) (C) (D) (E)
95. (A) (B) (C) (D) (E)
96. (A) (B) (C) (D) (E)
97. (A) (B) (C) (D) (E)
98. (A) (B) (C) (D) (E)
99. (A) (B) (C) (D) (E)
100. (A) (B) (C) (D) (E)

BIOLOGY EXAM V

PART A

DIRECTIONS: For each of the following questions or incomplete sentences, there are five choices. Choose the answer which is most correct. Darken the corresponding space on your answer sheet.

1. Which one of the following taxonomic groups includes all of the others?

(A) family
(B) genus
(C) class
(D) species
(E) order

2. The enzyme used in building a DNA molecule using an RNA template is

(A) restriction endonuclease
(B) DNA ligase
(C) reverse transcriptase
(D) RNA polymerase
(E) deoxyribonuclease

3. The group that includes members that have RNA as their genetic material is the

(A) bacteria
(B) fungi
(C) protozoa
(D) viruses
(E) insects

4. Two unrelated organisms that become similar in appearance and ways of life as they adapt to similar environmental situations exhibit

(A) adaptive radiation
(B) convergent evolution
(C) divergent evolution
(D) homology
(E) parallel evolution

5. Thylakoids occur in

(A) centrioles
(B) chloroplasts
(C) lysosomes
(D) mitochondria
(E) ribosomes

6. The final step in the clotting of blood involves the conversion of

(A) prothrombin to thromboplastin
(B) thrombin to thromboplastin
(C) fibrin to fibrinogen
(D) fibrinogen to fibrin
(E) prothrombin to thrombin

7. Insect tracheae differ from gills and lungs in that they

(A) only transport oxygen

(B) only transport carbon dioxide

(C) provide outpocketed gas exchange surfaces

(D) are not associated with a circulatory system

(E) are held rigid by an inner lining of cartilage

8. In flowering plants, the process that enables the sperm to approach the egg is

(A) fertilization

(B) growth of the pollen tube

(C) motility of the sperm

(D) pollination

(E) B and D

9. An XXX individual

(A) is a male

(B) has Down's syndrome

(C) has three Barr bodies

(D) has an X-linked trait

(E) results following nondisjunction

10. In many mammals this membrane, which contains both fetal and maternal tissue, serves as a pathway for transferring food, oxygen and other substances to the embryo and wastes to the mother.

(A) allantois
(B) amnion
(C) cutaneous
(D) placenta
(E) yolk sac

11. A cell formed by meiosis contains 12 chromosomes. The cell that divided to form it contained

(A) 6 chromosomes
(B) 12 chromosomes
(C) 24 chromosomes
(D) 36 chromosomes
(E) 48 chromosomes

12. Which of the following are found in equal proportions in DNA?

(A) adenine and cytosine
(B) adenine and uracil
(C) adenine and thymine
(D) cytosine and thymine
(E) thymine and guanine

13. In the scientific name *Escherichia coli*, *Escherichia* is the

(A) phylum
(B) class
(C) genus
(D) family
(E) order

14. Vitamins are required by animals and many microorganisms for use as

(A) energy sources

(B) structural materials

(C) parts of enzyme systems

(D) mineral sources

(E) parts of hormones

15. According to the heterotroph hypothesis, which gas was not present in the early atmosphere as the organic molecules that gave rise to the earliest life forms on earth were forming?

(A) ammonia (NH_3)

(B) hydrogen (H_2)

(C) methane (CH_4)

(D) oxygen (O_2)

(E) water vapor (H_2O)

16. Which of the following extinct forms most closely resembles present day *Homo sapiens*?

(A) *Australopithecus*

(B) *Homo erectus*

(C) *Homo habilis*

(D) Cro-Magnon

(E) *Zinjanthropus*

17. The Hardy-Weinberg equilibrium would not prevail under which of the following conditions?

(A) absence of selection

(B) large population

(C) presence of random mating

(D) presence of mutation

(E) absence of immigration or emigration

18. Which of the following structures develops from embryonic ectoderm?

(A) brain

(B) circulatory system

(C) lining of digestive tube

(D) liver

(E) skeleton

19. Reproduction in which an unfertilized egg develops into an adult animal is termed

(A) conjugation

(B) hermaphroditism

(C) metamorphosis

(D) parthenogenesis

(E) regeneration

20. The form of learning exhibited by fish that swim to the surface in response to the turning-on of a light when they originally responded to the presence of food provided at the same time a light was lit is known as

(A) conditioning
(B) habituation
(C) imprinting
(D) insight
(E) instinct

21. As compared to the success of other vascular plants and bryophytes (liverworts and mosses), how could you account for the success and predominance of flowering plants and conifers on land?

(A) they have roots, stem and leaves
(B) they use chlorophyll-a as a light-trapping pigment in photosynthesis
(C) they form large conspicuous gametophytes
(D) exposed parts are covered by a waxy cuticle
(E) the evolution of pollen and seeds eliminated the need for external water for fertilization

22. An animal hormone that causes the heart to beat more rapidly is

(A) epinephrine
(B) estrogen
(C) insulin
(D) oxytocin
(E) parathyroid hormone

23. A tissue or tissues that lack blood vessels and must receive nutrients by diffusion from a neighboring tissue is (are):

1. cartilage
2. epithelium
3. muscular tissue

(A) 1 only
(B) 2 only
(C) 3 only
(D) 1 and 2
(E) 1 and 3

24. How does meiosis differ from mitosis?

1. it contributes to variability
2. it forms monoploid cells from diploid cells
3. it forms genetically identical daughter cells

(A) 1 only
(B) 2 only
(C) 3 only
(D) 1 and 2
(E) 1 and 3

25. This process accounts for the ability of the freshwater alga, *Nitella*, to accumulate a concentration of potassium ions more than a thousand times greater than that of the surrounding water.

(A) active transport
(B) osmosis
(C) phagocytosis
(D) simple diffusion
(E) facilitated diffusion

26. The acetyl (two-carbon) group is to oxaloacetic acid in the Krebs cycle as CO_2 is to what compound in the dark reactions (Calvin cycle) of photosynthesis?

(A) glucose

(B) phosphoglyceric acid (PGA)

(C) phosphoglyceraldehyde (PGAL)

(D) pyruvic acid

(E) ribulose bisphosphate

27. Which of the following are characteristic of normal mature human red blood cells?

1. they have nuclei
2. they contain hemoglobin
3. they are capable of amoeboid movement

(A) 1 only

(B) 2 only

(C) 3 only

(D) 1 and 2

(E) 1, 2 and 3

28. Under certain circumstances, gallstones may block the passage of what digestive juice or juices?

1. bile
2. gastric juice
3. pancreatic juice
4. saliva

(A) 2 only
(B) 3 only
(C) 4 only
(D) 1 and 2
(E) 1 and 3

29. The main structures used for gas exchange in dry air by active land animals are

1. body surfaces
2. gills
3. lungs
4. tracheal tubes

(A) 1 only
(B) 3 only
(C) 4 only
(D) 2 and 3
(E) 3 and 4

30. All of the following are examples of homeostatic mechanisms EXCEPT

(A) after exercising vigorously you perspire and the evaporation of water from the skin lowers body temperature
(B) phagocytic cells in a rabbit detect and eat bacteria
(C) a frog deposits its eggs in a pond
(D) shivering occurs outdoors when the air temperature is low
(E) a small cut in a finger triggers the clotting reaction

31. The correct sequence by which air on its way to the lungs passes through the following passages is

(A) nasal cavity-pharynx-larynx-bronchus-trachea

(B) nasal cavity-pharynx-larynx-trachea-bronchus

(C) nasal cavity-pharynx-trachea-larynx-bronchus

(D) nasal cavity-larynx-pharynx-bronchus-trachea

(E) nasal cavity-larynx-pharynx-trachea-bronchus

32. Which answer below lists the organisms in a sequence that corresponds to the following?

intracellular digestion only → digestion in an unbranched cavity that has one opening → digestion partly in a branched cavity that has one opening → digestion in a tube that has two openings

1. *Hydra*
2. earthworm
3. sponge
4. planarian

(A) 1-2-3-4

(B) 3-1-4-2

(C) 3-4-1-2

(D) 1-4-3-2

(E) 1-3-4-2

33. Flowering plants and conifers have all of the following in common, EXCEPT

(A) vascular tissue for conduction and support

(B) pollen

(C) seeds

(D) conspicuous sporophytes

(E) fruits

34. The abnormal hemoglobin of individuals with sickle cell anemia is a result of nucleotide

1. addition
2. deletion
3. substitution

(A) 1 only
(B) 2 only
(C) 3 only
(D) 1 and 3
(E) 2 and 3

35. In comparing photosynthesis and cellular respiration which one of the following statements would not be true?

(A) ATP is formed during both processes
(B) CO_2 is produced during both processes
(C) O_2 is released during photosynthesis only
(D) several enzymes are needed for each process to occur
(E) both processes occur in green plants

36. A classification scheme that places pine trees, tomato plants, and pepper plants in one group and ferns and liverworts in another group is based on

1. vascular versus nonvascular
2. flowering versus nonflowering
3. seed forming versus nonseed forming

(A) 1 only
(B) 2 only
(C) 3 only
(D) 1 and 2
(E) 1 and 3

37. How could you account for the presence of coniferous forest in West Virginia or California, since such forests characteristically form a broad band across North America, Europe and Asia at latitudes to the north of that of the Great Lakes?

(A) they occur only in areas of very low rainfall

(B) they occur only in areas having similar animals

(C) they occur only in areas with colder temperatures at higher elevations on mountains

(D) they occur only in areas of permafrost

(E) they occur only in areas having well-drained soil

38. The classification of a perch, a frog, and a rabbit in the same group is based on

(A) being warm-blooded

(B) using lungs for gas exchange

(C) using gills for gas exchange

(D) having backbones of vertebrae

(E) having four-chambered hearts

39. Which one of the following shows a correct sequence of energy flow in an ecosystem?

1. carnivore
2. decomposer
3. herbivore
4. producer

(A) 4-1-3-2
(B) 2-4-1
(C) 1-3-2
(D) 4-2
(E) 1-2-3-4

40. Over time a single population gives rise to two animal species. The most likely sequence of events would be

1. reproductive isolation
2. variation
3. natural selection
4. geographic isolation

(A) 1-2-3-4
(B) 3-1-2-4
(C) 4-1-2-3
(D) 4-2-3-1
(E) 4-1-3-2

41. A son with hemophilia could result from which of the following crosses?

1. X^HX^H x X^HY
2. X^HX^H x X^hY
3. X^HX^h x X^hY

(A) 1 only
(B) 2 only
(C) 3 only
(D) 1 and 2
(E) 2 and 3

42. All of the following are not true of mutations that occur in humans, EXCEPT

(A) only mutations that are dominant are passed to the offspring

(B) none of the mutations are passed to offspring

(C) only mutations occurring in the cells that form gametes are passed to offspring

(D) only mutations that are recessive are passed to the offspring

(E) all mutations occurring in parents are passed to their offspring

43. What characteristic(s) does an oak tree have in common with a mushroom?

(A) utilization of chlorophyll

(B) possession of eukaryotic cells

(C) heterotrophic capabilities

(D) utilization of chlorophyll and possession of eukaryotic cells

(E) possession of eukaryotic cells and heterotrophic capabilities

44. All of the following contain DNA EXCEPT

(A) chloroplast

(B) mitochondrion

(C) nucleus

(D) plasmid

(E) ribosome

45. The element nitrogen is present in all compounds of which of the following classes of organic molecules?

(A) carbohydrates and nucelic acids

(B) lipids and nucleic acids

(C) carbohydrates and lipids

(D) nucleic acids and proteins

(E) carbohydrates and proteins

46. Which of the following is not characteristic of enzymes?

(A) they increase the rate of a chemical reaction

(B) they can be denatured

(C) they lower the requirement of a reaction for activation energy

(D) they bind with specific substrate molecules

(E) they are used up in the chemical reaction

47. Which of the following is not present in prokaryotic cells?

(A) chromosome

(B) DNA

(C) flagellum

(D) mitochondrion

(E) ribosome

48. Eukaryotic cells that produce and export large quantities of proteins, such as liver cells, would be well supplied with which organelle?

(A) centriole

(B) lysosome

(C) mictotubule

(D) mitochondrion

(E) rough endoplasmic reticulum

49. Given the genetic code below and the sequence of bases on a sense strand of DNA is 3' TAC-AAA-GAC-TAT 5', what would the last amino acid be in the translated amino acid chain?

(A) methionine
(B) tyrosine
(C) isoleucine
(D) leucine
(E) phenylalanine

The Genetic Code

(Based on Messenger RNA Codons)

First Base	Second Base: U	C	A	G	Third Base
U	Phenylalanine	Serine	Tyrosine	Cysteine	U
	Phenylalanine	Serine	Tyrosine	Cysteine	C
	Leucine	Serine	Stop	Stop	A
	Leucine	Serine	Stop	Tryptophan	G
C	Leucine	Proline	Histidine	Arginine	U
	Leucine	Proline	Histidine	Arginine	C
	Leucine	Proline	Glutamine	Arginine	A
	Leucine	Proline	Glutamine	Arginine	G
A	Isoleucine	Threonine	Asparagine	Serine	U
	Isoleucine	Threonine	Asparagine	Serine	C
	Isoleucine	Threonine	Lysine	Arginine	A
	start Methionine	Threonine	Lysine	Arginine	G
G	Valine	Alanine	Aspartic acid	Glycine	U
	Valine	Alanine	Aspartic acid	Glycine	C
	Valine	Alanine	Glutamic acid	Glycine	A
	Valine	Alanine	Glutamic acid	Glycine	G

50. Given the amino acid inserted in a protein is alanine, which of the following could be the DNA triplet that coded for it?

(A) CGU
(B) CAA
(C) CTA
(D) CAT
(E) CGA

51. In addition to pyruvate and ATP, what is the other end-product of glycolysis?

(A) O_2
(B) CO_2
(C) $NADH_2$ ($NADH + H^+$)
(D) $FADH_2$
(E) $NADPH_2$ ($NADPH + H^+$)

52. In which kingdom are organisms with prokaryotic cells classified, if the five-kingdom system of classification is used?

(A) animalia
(B) fungi
(C) monera
(D) plantae
(E) protista

53. If the three-nucleotide sequence on one strand of DNA is 3' TAC 5', what is the three-nucleotide sequence on its DNA complement?

(A) 5' AUG 3'
(B) 5' ATG 3'
(C) 3' AUG 5'
(D) 3' TAG 5'
(E) 5' CGA 3'

54. An *Amoeba* lives on the underside of a lily pad and feeds on bacteria. The lily pad is its:

(A) niche
(B) habitat
(C) biome
(D) biosphere
(E) ecosystem

55. The correct sequence of the following in the lytic cycle is

1. synthesis of phage proteins and nucleic acids
2. assembly of complete phages
3. injection of phage nucleic acids
4. lysis of host bacterium and release of complete phages
5. attachment of phage to host bacterium

(A) 1-2-3-4-5
(B) 5-3-1-2-4
(C) 5-1-2-3-4
(D) 5-3-2-1-4
(E) 5-2-3-1-4

56. Organisms capable of nitrogen fixation are in which of these groups?

(A) monera
(B) diatoms
(C) legumes
(D) protozoa
(E) rotifers

57. A substance secreted by an individual that affects the behavior of other members of the same species is a

(A) histone
(B) hormone
(C) pheromone
(D) prostaglandin
(E) steroid

Base Composition in Mole Percent

Organism	Base			
	Adenine	Thymine	Guanine	Cytosine
Human	30.9	29.4	19.9	19.8
Locust	29.3	29.3	20.5	20.7
Wheat	27.3	27.1	22.7	22.8
Hobbit	30.1	-----	-----	-----

58. Given the information above, what is the approximate percentage of guanine in a hobbit?

(A) 10
(B) 20
(C) 30
(D) 40
(E) 60

PART B

DIRECTIONS: Each of the following sets of questions refers to a list of lettered choices immediately preceding the question set. Choose the best answer from the list, then darken the corresponding space on your answer sheet. Choices may be used once, more than once, or not at all.

Questions 59 - 61

(A) arthropoda
(B) cnidaria
(C) echinodermata
(D) mollusca
(E) porifera

59. Phylum of segmented animals that have jointed appendages and exoskeletons of chitin

60. Radially symmetrical animals that are characterized by mouths surrounded by tentacles bearing nematocysts are in this phylum

61. Animals in this group are bilaterally symmetrical and have mantles; these mantles may secrete shells that cover the organism's bodies

Questions 62 - 64

(A) commensal
(B) parasite
(C) predator
(D) saprophyte
(E) scavenger

62. A fox catches and eats a mouse.

63. A squirrel lives in a nest it built in a tulip poplar tree.

64. A worm (hookworm) bores through the skin of a human, migrates to the intestine, attaches and feeds on blood.

Questions 65 - 66

(A) amphibians
(B) birds
(C) bony fish
(D) mammals
(E) reptiles

65. Adults use gills covered by an operculum for gas exchange

66. Terrestrial adults produce eggs that must be deposited in water

Questions 67 - 69

(A) gall bladder
(B) kidney
(C) liver
(D) pancreas
(E) spleen

67. This organ produces insulin and glucagon, which help to regulate blood glucose level

68. the main organ involved in forming the nitrogen-containing waste urea

69. the main organ involved in the removal of urea from the blood

Questions 70 - 72

(A) apical meristem
(B) pericycle
(C) phloem
(D) vascular cambium
(E) xylem

70. Tissue that gives rise to secondary (lateral) roots

71. The main tissue involved in transporting water from roots to leaves

72. Growth in height of a tree results from the addition of cells in this area

Questions 73 - 75

(A) anaphase
(B) interphase
(C) metaphase
(D) prophase
(E) telophase

73. Chromosomes are oriented across the equator of the cell

74. DNA replication occurs during this phase of the cell cycle

75. Centromeres split and single chromosomes move to opposite poles of the spindle

Questions 76 - 78

(A) biome

(B) biosphere

(C) community

(D) ecosystem

(E) population

76. A group of similar individuals living in one particular area at a given time

77. Portion of the earth inhabitated by living organisms

78. All of the different kinds of organisms that live and interact in one place at a given time

PART C

DIRECTIONS: The following questions refer to the accompanying diagram. Certain parts of the diagram have been labelled with numbers. Each question has five choices. Choose the best answer and darken the corresponding space on your answer sheet.

Questions 79 - 81 refer to the diagram below:

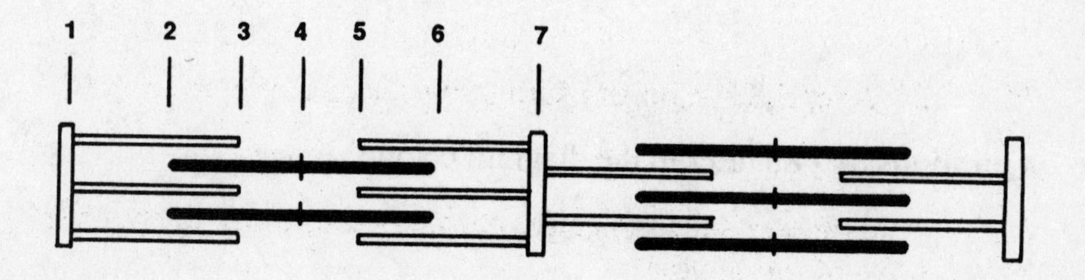

79. Which number indicates the location of a Z-line in the diagram above?

(A) 1
(B) 2
(C) 3
(D) 4
(E) 6

80. Which one of the following is not true about the area between numbers 2 and 6?

(A) it contains filaments with cross-bridges
(B) it remains the same length during contraction
(C) it is a complete sarcomere
(D) it corresponds to the location of thick filaments
(E) it is an A-band

81. The light area between numbers 1 and 2 is an:

(A) H-zone and contains thick filaments only
(B) H-zone and contains thin filaments only
(C) I-band and contains thick filaments only
(D) I-band and contains thin filaments only
(E) I-band containing overlapping thick and thin filaments

Questions 82 - 83 refer to the diagram below:

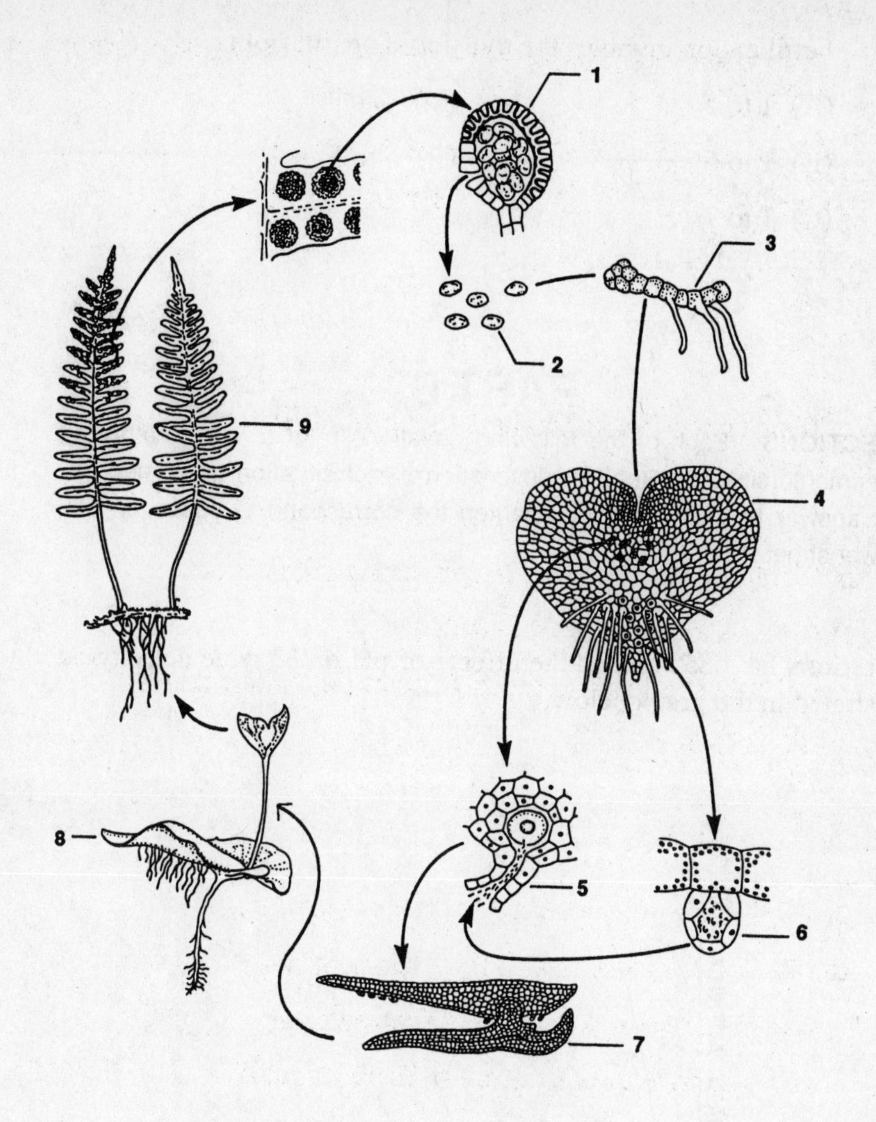

82. All of the following numbers designate haploid structures, EXCEPT

(A) 2
(B) 3
(C) 6
(D) 8
(E) 9

83. Fertilization involves the transfer of sperm from

(A) 1 to 5
(B) 1 to 6
(C) 1 to 7
(D) 6 to 7
(E) 6 to 5

PART D

DIRECTIONS: Each of the following questions refer to a laboratory or experimental situation. Read the description of each situation, then select the best answer to each question. Darken the corresponding space on the answer sheet.

Questions 84 - 85 refer to the effects of pH on enzyme activity as illustrated in the graph below.

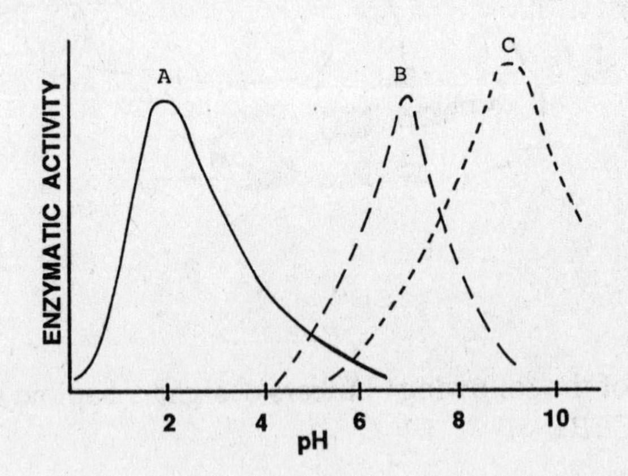

84. At which one of the following pHs are enzymes A, B and C functioning simultaneously?

(A) 4 (D) 7

(B) 4.5 (E) 7.5

(C) 5.8

85. Which of the following is true at pH 8?

(A) activity of enzyme A is decreasing

(B) activity of enzyme B is increasing

(C) activity of enzyme A is increasing

(D) activity of enzyme C is increasing

(E) activities of enzymes B and C are increasing

Questions 86 and 87 refer to the results of a blood typing test done on five individuals.

ABO Blood Group

Agglutination Results

Individual	Antiserum	
	Anti-A	Anti-B
1	+	+
2	-	+
3	-	-
4	+	-
5	-	+

86. From the results shown above, which individual has blood type AB?

(A) 1
(B) 2
(C) 3
(D) 4
(E) 5

87. Which individual above has the genotype ii?

(A) 1
(B) 2
(C) 3
(D) 4
(E) 5

Questions 88 - 90 refer to the breeding of curly-haired and straight-haired hobbits. The results are from five different crosses.

	Parents		Offspring	
Cross	Female	Male	Curly	Straight
1	Curly	Straight	0	58
2	Straight	Curly	48	52
3	Straight	Straight	38	119
4	Curly	Curly	48	0
5	Straight	Curly	0	49

88. If two curly-haired hobbits resulting from cross 3 mate with each other, what is the probability that their offspring will have straight hair?

(A) 3/4

(B) 1/2

(C) 1/4

(D) 1/8

(E) 0

89. If the breeder wanted to obtain homozygous straight-haired hobbits only, he would select which of the following as parents?

	Female of		Male of
(A) a female of	cross 2	and a male of	cross 1
(B) a female of	cross 3	and a male of	cross 1
(C) a female of	cross 2	and a male of	cross 3
(D) a female of	cross 3	and a male of	cross 3
(E) a female of	cross 5	and a male of	cross 1

90. If two curly-haired individuals are mated, what is the probability that a curly-haired female will be produced?

(A) 1/16

(B) 1/8

(C) 1/4

(D) 3/8

(E) 1/2

Questions 91 - 93

The four graphs below illustrate the influence of various environmental factors on the rate of photosynthesis.

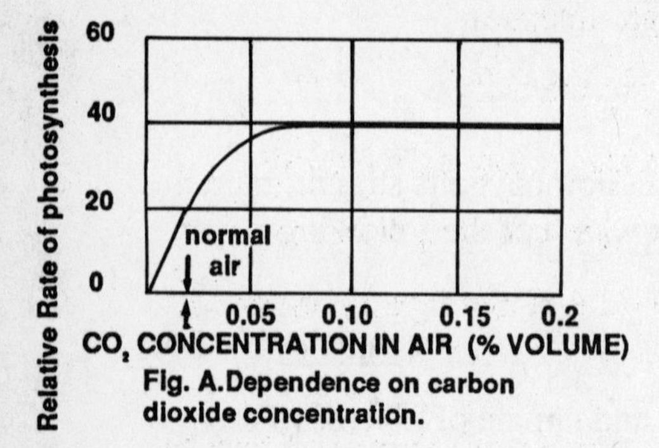

Fig. A. Dependence on carbon dioxide concentration.

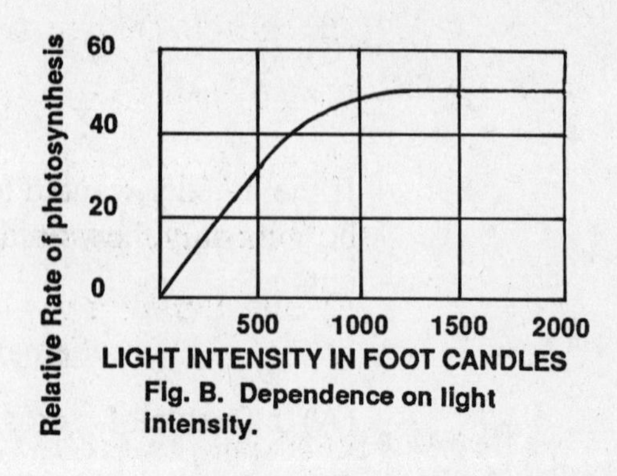

Fig. B. Dependence on light intensity.

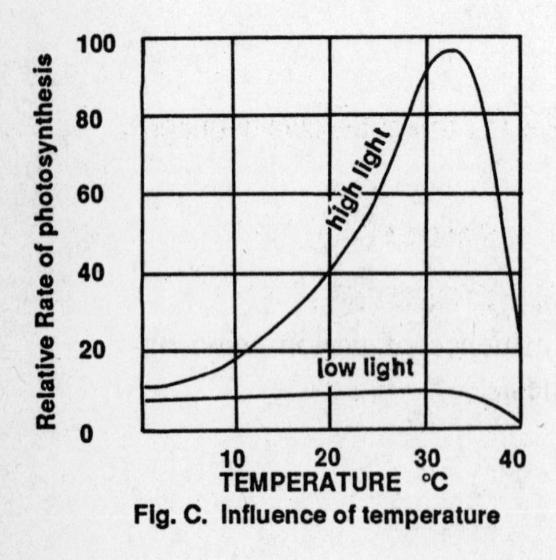

Fig. C. Influence of temperature

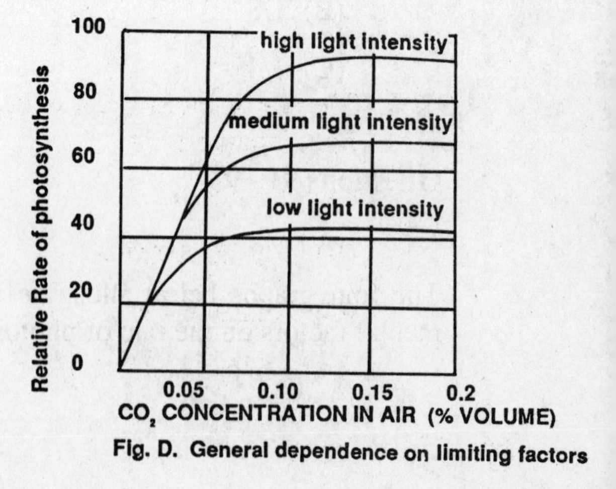

Fig. D. General dependence on limiting factors

91. At 30°C, the rate of photosynthesis at high light intensity compared to the rate at low light intensity is:

(A) equal to the rate at low light intensity

(B) less than the rate at low light intensity

(C) two times greater than the rate at low light intensity

(D) four times greater than the rate at low light intensity

(E) nine times greater than the rate at low light intensity

92. Which one of the following statements is not supported by data in the above figures?

(A) photosynthetic rate increases as the concentration of chlorophyll increases

(B) as the concentration of CO_2 increases, photosynthetic rate increases up to a limit

(C) photosynthetic activity decreases at temperatures above 35°C

(D) when enough CO_2 is available, light is a limiting factor in photosynthesis

(E) as the light intensity increases, photosynthetic activity increases up to a limit

93. The rate of photosynthesis increases in a linear fashion between light intensities of:

(A) 0 and 500

(B) 500 and 1000

(C) 1000 and 1500

(D) 1500 and 2000

(E) 2000 and 2500

Questions 94 - 96

Biome		J	F	M	A	M	J	J	A	S	O	N	D
A	T	1.1	1.7	6.1	12.2	17.8	22.2	25.0	23.3	20.0	13.9	7.8	2.2
	P	8.1	7.6	8.9	8.4	9.2	9.9	11.2	10.2	7.9	7.9	6.4	7.9
B	T	25.6	25.6	24.4	25.0	24.4	23.3	23.3	24.4	24.4	25.0	25.6	25.6
	P	25.8	24.9	31.0	16.5	25.4	18.8	16.8	11.7	22.1	18.3	21.3	29.2
C	T	12.8	15.0	18.3	21.1	25.0	29.4	32.8	32.2	28.9	22.2	16.1	13.3
	P	1.0	1.3	1.0	0.3	0.0	0.0	0.3	1.3	0.5	0.5	0.8	1.0
D	T	–20.0	–18.9	–12.2	–2.2	5.6	12.2	16.1	15.0	10.6	3.9	–5.6	–15.0
	P	3.3	2.3	2.8	2.5	4.6	5.6	6.1	8.4	7.4	4.6	2.8	2.8
E	T	24.6	25.1	26.4	28.5	30.6	31.9	31.1	30.3	31.1	28.8	26.5	25.1
	P	0.8	0.5	1.3	0.5	0.3	0.3	0.0	0.3	0.3	0.3	0.3	0.3

Monthly variations in temperature (T) in degrees celsius and precipitation (P) in centimeters from five major land biomes of the earth are listed above as A, B, C, D and E.

94. Which data set above (A, B, C, D or E) is consistent with a tropical desert?

95. The climatogram graphed below represents which data set (A, B, C, D or E)?

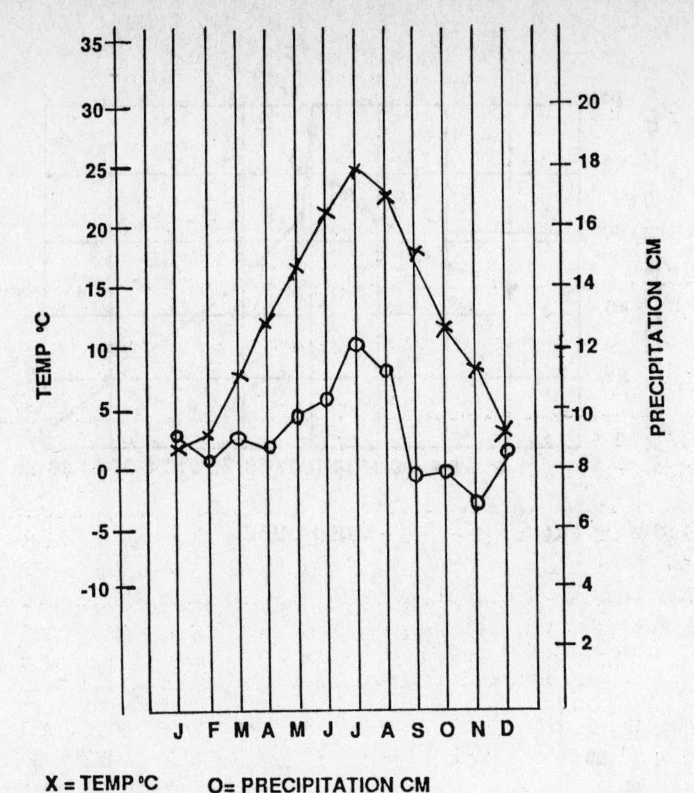

96. The temperature and precipitation data graphed in #95 above is most consistent with which one of the following land biomes?

(A) tropical rain forest

(B) mid-latitude desert

(C) tropical desert

(D) northern coniferous forest

(E) mid-latitude deciduous forest

Questions 97 - 98

In the early 1950s, Eckhard H. Hess and A. O. Ramsey studied the imprinting of mallard ducklings at various ages after hatching and at various distances travelled. Figure A shows the percent of positive responses versus critical age and Figure B shows the percentage of positive responses versus distance travelled.

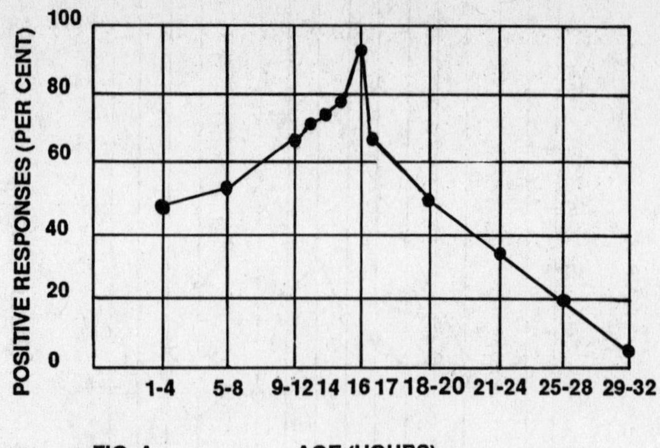

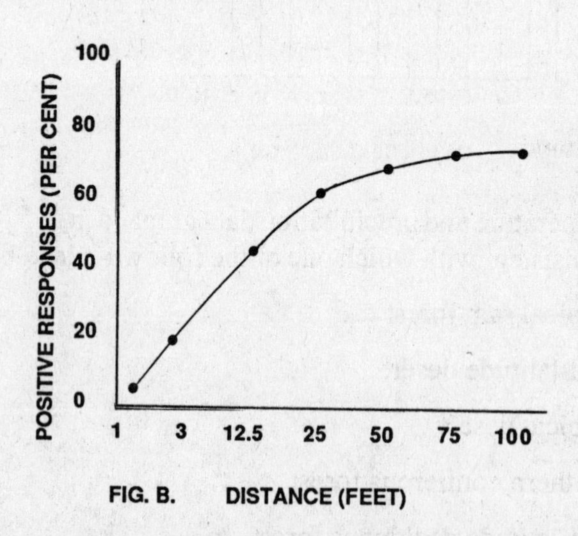

97. The data above indicates that ducklings are most sensitive to imprinting between:

(A) 1-4 hours after hatching

(B) 5-8 hours after hatching

(C) 13-16 hours after hatching

(D) 17-20 hours after hatching

(E) 41-44 hours after hatching

98. The greatest percentage of increase in imprinting in relation to distance travelled occurred between:

(A) 1 and 3 feet

(B) 3 and 12.5 feet

(C) 12.5 and 25 feet

(D) 25 and 50 feet

(E) 50 and 100 feet

Questions 99 - 100

Two sets of test tubes were set up as shown below. All of the tubes contained water and a few drops of the pH indicator bromthymol blue. The indicator is blue when the pH is above 7.5 and yellow when the pH is less than 6.0. All of the tubes are yellow at the start of the experiment. Tubes II and IV contain snails of similar size. Tubes III and VI contain comparable sprigs of the aquatic plant *Elodea*. Tubes I-III are kept in the light; tubes IV-VI sit in the dark. When the tubes are examined 12 hours later, the indicator in tube III is blue and indicator in the remaining tubes is still yellow.

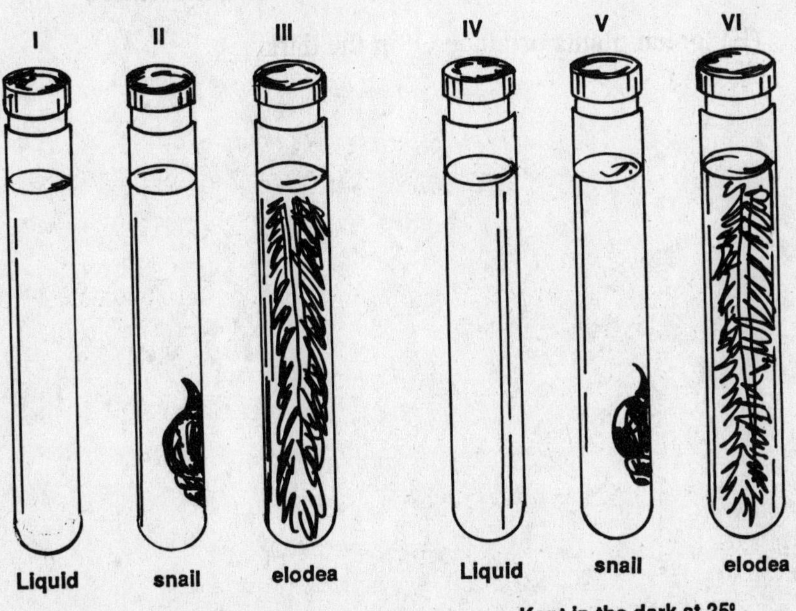

99. After 12 hours the indicator in Tube III is blue because:

(A) *Elodea* respired and gave off CO_2 which formed carbonic acid and lowered the pH

(B) indicator decomposed in the light causing the medium to become more basic

(C) *Elodea* photosynthesized and gave off O_2 which lowers the pH

(D) *Elodea* photosynthesized and used up CO_2 which caused the pH to become more basic

(E) the snail formed wastes which raised the pH

100. Which one of the following statements is not supported by the experiment?

(A) *Elodea* requires light for photosynthesis

(B) snails do not photosynthesize in the light or dark

(C) the pH indicator is not altered by the light

(D) the pH indicator is not altered by the temperature

(E) green plants produce O_2 in the dark

BIOLOGY
TEST V

ANSWER KEY

1.	C	34.	C	67.	D
2.	C	35.	B	68.	C
3.	D	36.	C	69.	B
4.	B	37.	C	70.	B
5.	B	38.	D	71.	E
6.	D	39.	D	72.	A
7.	D	40.	D	73.	C
8.	E	41.	C	74.	B
9.	E	42.	C	75.	A
10.	D	43.	B	76.	E
11.	C	44.	E	77.	B
12.	C	45.	D	78.	C
13.	C	46.	E	79.	A
14.	C	47.	D	80.	C
15.	D	48.	E	81.	D
16.	D	49.	C	82.	E
17.	D	50.	E	83.	E
18.	A	51.	C	84.	C
19.	D	52.	C	85.	D
20.	A	53.	B	86.	A
21.	E	54.	B	87.	C
22.	A	55.	B	88.	E
23.	D	56.	A	89.	E
24.	D	57.	C	90.	E
25.	A	58.	B	91.	E
26.	E	59.	A	92.	A
27.	B	60.	B	93.	A
28.	E	61.	D	94.	E
29.	E	62.	C	95.	A
30.	C	63.	A	96.	E
31.	B	64.	B	97.	C
32.	B	65.	C	98.	B
33.	E	66.	A	99.	D
				100.	E

BIOLOGY TEST V

DETAILED EXPLANATIONS OF ANSWERS

1. (C)

The correct sequence of taxonomic categories from largest to smallest is Kingdom, Phylum, Class, Order, Family, Genus, and Species.

2. (C)

Restriction endonucleases are bacterial enzymes that split DNA into pieces with specific nucleotides at their ends. In nature, the restriction enzymes are thought to protect bacteria from invading DNA viruses (bacteriophages or phages) by cleaving their DNA into useless fragments. In recombinant DNA technology, humans make use of over 250 restriction endonucleases for introducing DNA of one organism into the cells of another. DNA ligase is used to join short pieces of DNA to form a strand during DNA replication. Prior to that, a DNA polymerase is used to join DNA nucleotides. The enzyme deoxyribonuclease hydrolyzes (digests) DNA into nucleotides. RNA polymerase is used to join RNA nucleotides when RNA is transcribed from DNA. Reverse transcriptase, also called RNA-dependent DNA polymerase, is an enzyme that can build DNA using an RNA template.

3. (D)

Prokaryotes and eukaryotes are built of cells that contain both DNA and RNA. In them, DNA serves as the genetic material. DNA codes for RNAs, a process called transcription. In turn, one type of RNA, messenger RNA, codes for protein, a process called translation. Fungi, insects and protozoa have eukaryotic cells. Bacteria have prokaryotic cells. On the other hand, viruses are not cellular. They are infectious particles consisting of DNA or RNA (but not both), a protein coat, or capsid, and in some cases an envelope containing lipids and glycoproteins. The nucleic acid of a virus serves as its genetic material.

4. (B)

Divergent evolution and adaptive radiation are associated with increasing dissimilarity. The independent occurrence of similarities resulting from common ancestry is termed homology and the process is termed parallel evolution. Similar features (e.g.: wings of insects and birds) that occur independently in distantly related organisms in response to adapting to similar environments illustrate convergence.

5. (B)

Thylakoids are sac-like photosynthetic units located inside chloroplasts.

6. (D)

Fibrinogen is a soluble blood protein that, in the presence of thrombin, is converted into long threads of fibrin, forming the meshwork of a clot.

7. (D)

The open circulatory system of insects is apparently not involved in transporting respiratory gases. Tracheae strengthened by bands of chitin provide an extensive, inpocketed network of tubes that transport gases (O_2 and CO_2) to and from tissues.

8. (E)

Two processes enable the nonmotile sperm to approach the egg for fertilization to be accomplished. Firstly, pollen is transferred from the anther of a male stamen to the stigma of a female pistil, a process called pollination. Secondly, pollen germinates, forming a pollen tube that grows through the style to the egg inside the ovary.

9. (E)

Nondisjunction occurs when one or more chromosome pairs fail to separate normally during meiosis, resulting in the formation of gametes containing one or more extra chromosomes. If an egg containing 22 autosomes and two X-chromosomes is fertilized by a sperm that contains an X-chromosome, the resulting individual would be an XXX female. The number of Barr bodies in the cells of such an individual would be two (one less than the number of X-chromosomes).

10. (D)

In placental mammals, including humans, the placenta is an organ associated with the exchange of materials between mother and fetus. It develops from the embryonic chorion and maternal uterine tissue.

11. (C)

During meiosis (reduction division) the chromosome number is halved. If a cell formed by meiosis contains 12 chromosomes, then the cell that divided to form it must have contained 24 chromosomes.

12. (C)

Erwin Chargaff analyzed the base content of DNA in the cells of different species of eukaryotes and established that the amount of adenine equaled the amount of thymine and the amount of cytosine equaled the amount of guanine.

13. (C)

A scientific, or species, name is bipartite (e.g.: *Escherichia coli*). The first part of the name represents the genus (e.g. *Escherichia*) in which the species is classified and, when written, is capitalized and then italicized or underlined.

14. (C)

Vitamins form part of the structures of small organic molecules called coenzymes, which are essential to the functioning of certain enzymes.

15. (D)

The composition of the early atmosphere was an important factor in the origin and evolution of life on earth. Presumably, the early atmosphere was a reducing atmosphere, lacking molecular oxygen (O_2). Simple gases in it, including ammonia (NH_3), hydrogen (H_2), methane (CH_4), and water vapor (H_2O), contained the basic elements needed for the formation of life. Over time, in the absence of O_2, organic molecules formed and accumulated in the environment until the first forms of life evolved.

16. (D)

Cro-Magnon people, who were essentially like present day humans in appearance, first appeared (according to the fossil record) in Europe about 34,000 years ago, in Asia about 40,000 years ago and in Africa perhaps 100,000 years ago. *Homo erectus*, *Homo habilis* and *Australopithecus*, including *Zinjanthropus*, are much older; it is therefore thought that these species became extinct before the Cro-Magnons became dominant.

17. (D)

The Hardy-Weinberg equilibrium states that in a large population in which there is random mating, the frequency of alleles will remain constant in the absence of mutation, migration, and selection.

18. (A)

The embryonic brain develops from an inward migration of ectodermal cells. The circulatory system and skeleton develop from mesoderm and the lining of the digestive tube and the liver develop from endoderm.

19. (D)

The development of an egg without fertilization is termed parthenogenesis.

20. (A)

Conditioning is a form of learning in which a neutral stimulus, associated with a stimulus that normally elicits a behavioral response, causes the response in the absence of the usual stimulus. The fish involved have associated the neutral stimulus (light) with the concomitant introduction of food to their environment. Thus, they react upon seeing the light, even if they cannot yet detect the food.

21. (E)

Among the vascular plants and bryophytes, only the flowering plants and conifers have pollen and seeds. Two adaptations, pollination and growth of the pollen tube, enable flowering plants to accomplish fertilization in the absence of external water. Their seeds, which can remain dormant for extended periods of time, are adapted to protect and transport the embryonic plants contained in them.

22. (A)

Epinephrine (also called adrenalin) triggers the "fight-or-flight response." This response includes the breakdown of glycogen that is present in the liver and in muscle (whether the animal is to fight or run, it will still need a ready supply of energy, which is provided by glucose), and an increase in heart rate (the animal will need an increased supply of oxygen if it is to be able to utilize glucose).

23. (D)

Both cartilage and epithelium are avascular tissues that receive nutrients from neighboring tissues. Muscular tissue, however, is highly vascularized.

24. (D)

Meiosis, or reduction division, produces genetically different monoploid (= haploid) cells from diploid parent cells. The genetic variability results from the independent assortment of homologous chromosomes during meiosis I and from crossing-over.

25. (A)

Energy expended during active transport enables cells to accumulate substances despite the fact that these substances are in greater concentration inside the cell and will tend to diffuse out of the cell.

26. (E)

Just as Acetyl-coenzyme A is combined with oxaloacetic acid to initiate the Krebs cycle, the first step in the synthesis of carbohydrate by the Calvin cycle involves the combination of CO_2 and a molecule of the five-carbon sugar ribulose bisphosphate.

27. (B)

Erythrocytes, or red blood cells, which develop from nucleated cells in bone marrow, lose their nuclei as they enter the circulating blood. They contain the oxygen-transporting pigment hemoglobin and are incapable of independent movement such as that exhibited by amoeboid white blood cells.

28. (E)

Depending on where they lodge, in the bile ducts or in the pancreatic duct, gallstones can obstruct the flow of bile or pancreatic juice. Since these ducts empty into the small intestine, gallstones cannot affect the passage of gastric juice or of saliva.

29. (E)

For gas exchange on land, the majority of air breathing animals use inpocketed (invaginated) systems, lungs or tracheal tubes, kept moist by internal secretions. Some animals, such as frogs, also absorb oxygen through their skin, but they still depend on their lungs for the greater portion of their oxygen intake. Gills are simply too fragile and too susceptible to dessication to be of any use on land.

30. (C)

Homeostasis refers to the tendancy of an organism to maintain a more or less constant internal environment. The situations represented by all of the choices except C are ones in which an external or internal stress changes some normal condition to an abnormal condition. Homeostatic mechanisms serve to counteract the effects of any stresses on an organism. Therefore, when one exercises vigorously and one's internal body temperature increases, one perspires in order to remove excess internal heat, and to have one's internal body temperature return to normal.

31. (B)

The answer is self-explanatory.

32. (B)

Sponges depend on intracellular digestion exclusively. Both *Hydra* and planaria have digestive cavities with one opening. Such a digestive cavity is called a gastrovascular cavity. In planaria, the gastrovascular cavity is highly branched. Earthworms have a complete digestive tract with two openings, a mouth and an anus.

33. (E)

In reality, a fruit is the ripened ovary of a flower and may incorporate other closely associated flower parts. Conifers do not bear flowers, and so they also bear no fruit.

34. (C)

Sickle-cell hemoglobin (hemoglobin S) differs from normal adult hemoglobin (hemoglobin A) in that a single amino acid, valine, is substituted for the glutamic acid residue that is normally present as the sixth residue of each of the two B-subunits. This change in amino acid sequence is caused by the substitution of a single nucleotide in the DNA triplet that codes for glutamic acid.

35. (B)

CO_2 is incorporated into carbohydrates during the Calvin cycle of photosynthesis. CO_2 is formed during cellular respiration.

36. (C)

Pines, tomato, pepper plants and ferns are vascular plants. Of the plants that are mentioned, only tomato and pepper plants are flowering plants. However, pine trees, tomato plants, and pepper plants all produce seeds. Ferns and liverworts produce spores.

37. (C)

Extensions or isolated segments of coniferous forest (taiga, boreal forest) occur on the slopes of mountains, where the temperatures at the higher altitudes approximate those at more northern latitudes.

38. (D)

They, a perch, a frog, and a rabbit, are all classified as vertebrates because of their backbones of vertebrae. Of the animals that are listed, only rabbits have four-chambered hearts and are warm-blooded. Both rabbits and adult frogs use lungs for gas exchange, while perch and frog tadpoles have gills.

39. (D)

Producers capture energy and make it available to members of other trophic levels. The producers may be eaten by herbivores (primary consumers), or may die and be acted upon by decomposers.

40. (D)

An ancestral population may be split into two or more subgroups by geographic isolation. If, because of isolation, gene flow between the subgroups is prevented, mutations, natural selection and other factors cause the subgroups to become increasingly different genetically. Over time they may become so different genetically that interbreeding is not possible.

41. (C)

The gene for hemophilia is recessive and x-linked, thus, a son with hemophilia (X^hY) can only result from a mating in which the mother is either a carrier (X^HX^h) or a hemophiliac (X^hX^h). The father donates the Y-chromosome and so it does not matter if the father carries the gene for hemophilia or not. The Punnett squares show this below.

1.

	X^H	X^H
X^H	X^HX^H	X^HX^H
Y	X^HY	X^HY

2.

	X^H	X^H
X^h	X^HX^h	X^HX^h
Y	X^HY	X^HY

3.

	X^H	X^h
X^h	X^HX^h	X^hX^h
Y	X^HY	X^hY

42. (C)

In humans, only mutations occurring in the gametes (sperm and eggs) have the potential of being passed to offspring. These mutations can be either dominant or recessive, since such attributes do not affect whether or not a mutant gene will actually be present in a gamete; dominance and recessiveness only come into play in the determination of phenotype.

43. (B)

The oak tree, a flowering plant, contains chlorophyll and is autotrophic. The mushroom, a fungus, lacks chlorophyll and is heterotrophic. Both have eukaryotic cells.

44. (E)

Of the structures listed only ribosomes do not contain DNA. Ribosomes are composed of ribosomal RNA(rRNA) and associated proteins. A plasmid is a small, circular piece of DNA. The nucleus is the main repository for DNA in eukaryotic cells. However, both chloroplasts and mitochondria contain small circular DNA molecules that many biologists believe indicates their derivation from ancient prokaryotes.

45. (D)

All nucleic acids and proteins contain nitrogen in their chemical makeup. Nucleic acids contain nitrogenous bases (adenine, cytosine, guanine, thymine and uracil), and proteins are composed entirely of amino acids, each of which has an amino ($-NH_2$) group. Some amino acids also have nitrogen in their side chains.Some carbohydrates (e.g. chitin and bacterial cell wall polysaccharides) and some lipids (e.g. lecithin, a phospholipid) contain nitrogen, but many do not.

46. (E)

Enzymes increase the rates of chemical reactions by lowering the amount of activation energy that is needed. A factor that contributes to this lowering of activation energy is the fact that enzymes bind to reactants (substrates) and orient them in positions that are conducive to reactions. The enzymes then release the final products of the reaction and, unchanged, proceed to catalyze another reaction.

47. (D)

All prokaryotes have circular chromosomes of DNA and ribosomes of the 70 S type. Some prokaryotes move by using flagella, which differ structurally from eukaryotic flagella. However, mitochondria occur only in eukaryotic cells.

48. (E)

Proteins synthesized on ribosomes associated with endoplasmic reticula can enter the channels of the endoplasmic reticula and become enclosed in sacs; these sacs move to the membrane to export their contents to the outside.

49. (C)

The DNA triplet TAT codes for the mRNA codon AUA. The AUA codon results in the insertion of the amino acid isoleucine in the amino acid chain. You must remember the proper reading direction, which is from 5' to 3'. Thus you have the following:

3' TAC-AAA-GAC-TAT 5' (DNA sense strand)
5' AUG-UUU-CUG-AUA 3' (mRNA strand)

Reading the mRNA from 5' to 3', the last codon is AUA.

50. (E)

The DNA triplet CGA codes for the mRNA codon GCU which is translated into the amino acid alanine in a polypeptide. Choice (A) is incorrect because the base, uracil, is found in RNA, not DNA.

51. (C)

As phosphoglyceraldehyde (PGAL) is oxidized during glycolysis, oxidized nicotinamide adenine dinucleotide (NAD^+) is reduced, forming NADH+ H^+. O_2 is an electron acceptor in oxidative phosphorylation. CO_2 is produced in the conversion of pyruvate to acetyl coenzyme A and in the Krebs cycle. $FADH_2$ is produced in the Krebs cycle. $NADPH_2$ is produced by the pentose phosphate pathway.

52. (C)

Organisms with prokaryotic cells (i.e. bacteria and blue-green algae) are classified in Kingdom Monera. Organisms classified in the other four kingdoms have eukaryotic cells.

53. (B)

The two strands of a DNA double-helix run in opposite directions, 5' to 3' and 3' to 5'. 3' and 5' refer to the third and fifth carbons of deoxyribose. Wherever adenine occurs in one strand, thymine always appears in the complementary strand. Similarly, cytosine is always paired with guanine.

54. (B)

Within an ecosystem the place an organism lives (the lily pad) is its habitat and the role of the organism (bacteria eater) is its niche.

55. (B)

The answer is self-explanatory.

56. (A)

Nitrogen fixation (i.e. the incorporation of atmospheric nitrogen (N_2) into ammonia (NH_3)) is accomplished under anaerobic conditions by certain bacteria, cyanobacteria (blue-green algae). Legumes have a symbiotic relationship with certain Monerans, specifically bacteria. It is the bacteria that fix nitrogen rather than the legume itself.

57. (C)

Pheromones are compounds released into the environment by an individual to influence the behavior (i.e. attract a mate, mark a trail or territory) of other individuals of the same species. Histones are proteins that are associated with DNA. A hormone is a substance that is secreted by certain somatic cells that affects specific cells at other locations in the body. Prostaglandins are hormones that are fatty-acid derivatives. Steroids, many of which are hormones, all have structures that have three six-membered rings and one five-membered ring.

58. (B)

Since thymine is the complement of adenine and since there is just as much thymine in DNA as there is adenine, 30.1% of the bases are thymine. Therefore, 60.2% of the bases are adenine and thymine. The remaining 39.8% of the bases must include equal amounts of guanine and cytosine. Half of 39.8% is 19.9%, which is approximately 20%.

59. (A) 60. (B) 61. (D)

Characteristics of various inveretbrate phyla are described. The Echinoderms have a spiny skin and adult forms are radially symmetrical. The larval forms have bilateral symmetry. Phylum Porifera includes those organisms commonly known as sponges. These organisms have radial symmetry with pores scattered over the entire body.

62. (C) 63. (A) 64. (B)

A predator is an animal that captures, kills, and consumes another animal, its prey.

A commensal is an organism that benefits from its association with another type of organism, while the other organism neither benefits nor is harmed.

A parasite is an organism that perpetuates its existence by feeding on and harming another type of living organism, its host.

65. (C) 66. (A)

The characteristics of various vertebrate classes are described. Birds have feathers, four chambered hearts, and are warm-blooded. In addition, they lay hard shelled eggs which are protected from dessication (they don't need to lay eggs in water). Mammals have hair and mammary glands to nourish their young. They have four-chambered hearts and are warm-blooded. The reptiles are cold-blooded and have three-chambered hearts except crocodilions.. They lay leathery shelled eggs, protected from dessication on land.

67. (D)

The pancreas has a dual nature. Its exocrine portion secretes several digestive enzymes into the duodenum via a duct. Its endocrine portion, the Islets of Langerhans, secretes both insulin and glucagon, which help to regulate the glucose level of the blood.

68. (C)

An excess of amino acids in the diet leads to the formation of the nitrogenous waste urea ($H_2N\text{-}C(=O)\text{-}NH_2$) in the liver.

69. (B)

The kidneys are the principle locations at which urea is removed from the blood.

70. (B)

Secondary (lateral, or branch) roots arise by division of cells in the pericycle.

71. (E)

Xylem conducts water while phloem conducts sucrose and other material both up and down the plant body.

72. (A)

The increase in height of a tree results from the division and subsequent differentiation and elongation of cells located in apical meristems.

73. (C)

A eukaryotic cell cycle consists of the series of events that starts with the formation of a cell and ends with its division. Most of the cell cycle is spent in interphase preparing for division of the nucleus (mitosis), and division of the cytoplasm (cytokinesis). Mitosis consists of four phases involved in distributing identical sets of chromosomes to form two daughter nuclei identical to the original nucleus. The second phase of mitosis is metaphase, during which chromosomes and their copies (chromosomes and their copies are held together by centromeres) move to the equatorial plane of the cell.

74. (B)

During interphase, prior to the start of mitosis, DNA is replicated during the S (synthesis) stage.

75. (A)

Metaphase is followed by anaphase, during which the centromere of each chromosome splits, and identical sets of single chromosomes move to opposite poles of the spindle.

76. (E)

A population is a group of similar organisms living in one particular area at a given time.

77. (B)

The biosphere constitutes the portion of the earth (atmosphere, hydrospere, lithosphere) inhabitated by living organisms. A biome is a complex of ecological communities characterized by a distinctive type of vegetation, as determined by climate. Examples of biomes include: tundra, taiga, temperate deciduous forest, tropical rain forest, desert, and grassland.

An ecosystem is the biotic and abiotic factors of an ecological community considered together.

78. (C)

The biological community consists of all of the different kinds of organisms that live and interact in one place at a given time.

79. (A)

Z-lines are designated by numbers 1 and 7.

80. (C)

A sarcomere extends from one Z-line to the next. So, the area between numbers 1 and 7 depicts a sarcomere. The area from 2 to 6 is an A-band. It consists of thick myosin filaments alternating with thin actin filaments.

81. (D)

The area between 1 and 2 is an I-band. It contains only actin filaments which are the thin filaments. In the center of the A-band (area 2 to 6) is a lighter zone containing only thick filaments which are myosin. This zone is the H-zone and is indicated by the area between 3 and 5.

82. (E) 83. (E)

Meiosis, or reduction division, occurs inside fern sporangia, which are on fern leaves. The resulting haploid spores (#2) germinate to form haploid gametophytes (#'s 3, 4, and 8). Subsequently, a haploid sperm, formed in an antheridium (#6), and a haploid egg, formed in an archegonium (#5) located on the underside of the gametophyte, unite to form a diploid zygote. The diploid zygote develops into a diploid sporophyte (#'s 7 and 9).

84. (C)

Every enzyme has a minimum, optimum, and maximum pH. The pH at which the rate of the reaction catalyzed by an enzyme is greatest is called the optimum pH. At pHs below the minimum and above the maximum, enzyme activity is virtually zero. Enzymes B and C are inactive at pH 4. Enzyme C is inactive at pH 4.5. At pHs 7 and 7.5 enzyme A is inactive.

85. (D)

At pH=8, the activity of enzyme C is increasing, as indicated by the upward slope of the curve.

86. (A)

Antiserums Anti-A and Anti-B contain antibodies that react with antigens A and B, respectively, causing agglutination (clumping of red blood cells). Anti-A causes clumping of cells containing antigen A. Likewise, Anti-B causes clumping of cells containing antigen B. Since an individual with type AB has cells with both antigens, agglutination occurs in the presence of both Anti-A and Anti-B.

87. (C)

The simplest explanation of ABO blood group inheritance is based on three alleles: I^A, I^B and i. I^A codes for antigen A. I^B for antigen B. Neither antigen A or antigen B is coded for by i. I^A and I^B are codominant alleles, and both of them are dominant to allele i. Postulated genotypes for each blood group are:

Blood Group	Genotype(s)
A	I^AI^A or I^Ai
B	I^BI^B or I^Bi
AB	I^AI^B
O	ii

An individual with genotype ii has blood type O and produces neither antigen A or B.

88. (E)

Results indicate that the allele for straight hair is dominant to the allele for curly hair. Using "S" to represent the dominant allele for straight hair and "s" the recessive allele for curly hair, the cross can be interpreted as follows:

	Genotypes of Parents		**Offspring**
Cross	Female	Male	Phenotypic Ratio
1	ss	SS	All straight
2	Ss	ss	1/2 curly, 1/2 straight
3	Ss	Ss	1/4 curly, 3/4 straight
4	ss	ss	All curly
5	SS	ss	All straight

The two curly-haired hobbits referred to in question #88 would produce the same results as Cross 4 above.

89. (E)

In order to produce only progeny that are homozygous for the dominant allele for straight hair, both parents must also be homozygous for this allele. Of the listed parents, only female #5 and male #1 have this genotype.

90. (E)

If two curly-haired individuals are mated (ss x ss) the probability of producing an offspring with curly hair is 100% or 1. Sex inheritance is determined by sex chromosomes. Females, who have cells that contain two X-chromosomes, can only produce X eggs. Males, who have cells that contain one X-chromosome and one Y-chromosome, produce two types of sperm in equal numbers: X-containing and Y-containing. Since half of the sperm are X-containing and all of the eggs are X-containing, the chance of producing an XX female is 1/2. The chance of producing a female with curly hair is the product of the probability of producing a female (1/2) and the probability of producing an offspring with curly hair (100% or 1), which is 1/2.

91. (E)

At 30°C, Figure C shows that the relative rate of photosynthesis at low light intensity is approximately 10 and the rate at high light intensity is approximately 90 or nine times greater.

92. (A)

None of the figures takes into account the chlorophyll concentration as a factor in photosynthetic rate. Choice (B) is supported by the data shown in Figure A. The results in Figure C support the statement in choice (C). Choice (D) summarizes the results in Figure (D). The results in Figure B support the statement in choice (E).

93. (A)

Between light intensities (in foot candles) of 0 and 500, the rate of photosynthesis increases in a linear (straight line) fashion.

94. (E)

In a tropical desert precipitation is uniformly low and temperature uniformly high throughout most of the year.

95. (A)

If the data for each of the biomes were plotted, the resulting figure for biome A would resemble the figure provided in the question.

96. (E)

Temperature extremes are too low for the tropics but too high for a Northern coniferous forest. Yearly precipitation is too high for deserts. The data for biome B indicates a tropical rain forest. Biome C is a desert climate with high temperatures and low precipitation. A coniferous forest or taiga is indicated by the data in biome D.

97. (C)

The highest percentage of positive responses occurred at 13 to 16 hours after hatching.

98. (B)

The increase in imprinting in relation to distance travelled was about 30 percent for distances between 3 and 12.5 feet. The next highest percentage of increase was about 20 percent for distances between 12.5 and 25 feet.

99. (D)

In light of sufficient intensity the rate of photosynthesis by *Elodea* would be expected to exceed the rate of respiration resulting in a net decrease in the CO_2 content of the solution. As CO_2 is used in photosynthesis, carbonic acid in the water is converted to CO_2 (which is subsequently utilized) and H_2O resulting in an elevation of the pH of the solution and the blue color of tube III.

100. (E)

Green plants only produce O_2 in the light during photosynthesis.

BIOLOGY

TEST VI

BIOLOGY
ANSWER SHEET

1. A B C D E
2. A B C D E
3. A B C D E
4. A B C D E
5. A B C D E
6. A B C D E
7. A B C D E
8. A B C D E
9. A B C D E
10. A B C D E
11. A B C D E
12. A B C D E
13. A B C D E
14. A B C D E
15. A B C D E
16. A B C D E
17. A B C D E
18. A B C D E
19. A B C D E
20. A B C D E
21. A B C D E
22. A B C D E
23. A B C D E
24. A B C D E
25. A B C D E
26. A B C D E
27. A B C D E
28. A B C D E
29. A B C D E
30. A B C D E
31. A B C D E
32. A B C D E
33. A B C D E
34. A B C D E
35. A B C D E
36. A B C D E
37. A B C D E
38. A B C D E
39. A B C D E
40. A B C D E
41. A B C D E
42. A B C D E
43. A B C D E
44. A B C D E
45. A B C D E
46. A B C D E
47. A B C D E
48. A B C D E
49. A B C D E
50. A B C D E
51. A B C D E
52. A B C D E
53. A B C D E
54. A B C D E
55. A B C D E
56. A B C D E
57. A B C D E
58. A B C D E
59. A B C D E
60. A B C D E
61. A B C D E
62. A B C D E
63. A B C D E
64. A B C D E
65. A B C D E
66. A B C D E
67. A B C D E
68. A B C D E
69. A B C D E
70. A B C D E
71. A B C D E
72. A B C D E
73. A B C D E
74. A B C D E
75. A B C D E
76. A B C D E
77. A B C D E
78. A B C D E
79. A B C D E
80. A B C D E
81. A B C D E
82. A B C D E
83. A B C D E
84. A B C D E
85. A B C D E
86. A B C D E
87. A B C D E
88. A B C D E
89. A B C D E
90. A B C D E
91. A B C D E
92. A B C D E
93. A B C D E
94. A B C D E
95. A B C D E
96. A B C D E
97. A B C D E
98. A B C D E
99. A B C D E
100. A B C D E

BIOLOGY EXAM VI

PART A

DIRECTIONS: For each of the following questions or incomplete sentences, there are five choices. Choose the answer which is most correct. Darken the corresponding space on your answer sheet.

1. In the sense strand of a DNA double helix, thymine is transcribed to which of the following nitrogen bases in the complementary mRNA?

   (A) Adenine
   (B) Cytosine
   (C) Guanine
   (D) Thymine
   (E) Uracil

2. If a male is colorblind and the female is a carrier, what genotypes would be possible in their daughters?

   I. $X^C X^C$
   II. $X^C X^c$
   III. $X^c X^c$
   IV. $X^c Y$

   (A) II only
   (B) III only
   (C) IV only
   (D) II and III
   (E) I, II and III

3. The type of behavior exhibited by a moth that is attracted by and flies to a light is known as

(A) Conditioning
(B) Habituation
(C) Imprinting
(D) Learned
(E) Taxis

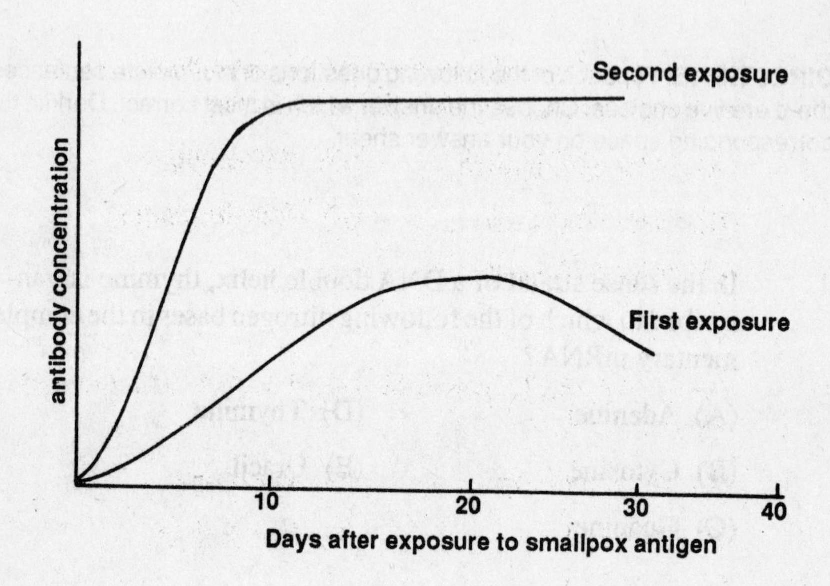

4. Based on the above graph, which of the following is true about exposure to the smallpox antigen?

(A) The antibody concentration increases faster in a first exposure.

(B) The concentration of antibody produced is greater in a first exposure.

(C) The antibody concentration peaks within 10 days after a first exposure.

(D) The concentration of antibody produced is greater after a second exposure.

(E) The concentration of antibody produced is the same after a first and a second exposure.

5. DNA and RNA are found in all of the following except individual

(A) Bacteria
(B) Diatoms
(C) Ferns
(D) Protozoa
(E) Viruses

6. The series of changes in form during development of an immature animal into an adult is known as

(A) Hermaphroditism
(B) Metamorphosis
(C) Parthenogenesis
(D) Transduction
(E) Transformation

7. The specialized branch of biology that deals with the study of birds is called:

(A) Helminthology
(B) Herpetology
(C) Ichthyology
(D) Oncology
(E) Ornithology

8. How does mitosis differ from meiosis?

I. mitosis contributes to variability
II. mitosis forms monoploid cells from diploid cells
III. mitosis forms genetically identical daughter cells

(A) I only
(B) II only
(C) III only
(D) I and II
(E) II and III

9. All of the following are associated with the light dependent reactions of photosynthesis except:

(A) Carbon dioxide
(B) Chlorophyll a
(C) Ferredoxin
(D) NADPH
(E) Plastoquinone

10. The 14 species of Darwin's finches on the Galapagos Islands are believed to have evolved from competing populations of an ancestral finch that colonized the islands from the South American mainland. If so, this would be an example of:

I. Adaptative radiation
II. Convergent evolution
III. Divergent evolution

(A) I only
(B) II only
(C) III only
(D) I and II
(E) I and III

11. What characteristic(s) does a yeast cell have in common with an elephant?

I. Chlorophyll
II. Eukaryotic
III. Heterotrophic

(A) I only
(B) II only
(C) III only
(D) I and II
(E) II and III

12. Which element is found in all proteins but not in all carbohydrates and lipids?

(A) Carbon
(B) Hydrogen
(C) Nitrogen
(D) Oxygen
(E) Phosphorus

13. Ferns and flowering plants have all of the following in common except:

(A) Eukaryotic cells

(B) Dominant sporophyte generation

(C) Free swimming sperm cells

(D) True roots, stems, and leaves

(E) Vascular tissues

14. The two-carbon acetyl group enters the Krebs Cycle by combining with which four-carbon molecule?

(A) Citrate

(B) Fumarate

(C) Oxaloacetate

(D) Pyruvate

(E) Succinate

15. A representative sample of a large population revealed that 1,600 out of the 10,000 individuals examined displayed a given recessive phenotype. Assuming the population is in Hardy-Weinberg equilibrium, how many individuals out of 100 would be expected to be homozygous dominant for the trait?

(A) 16

(B) 24

(C) 36

(D) 48

(E) 84

16. Which of the following are characteristic of mature human white blood cells?

I. Have nuclei
II. Contain hemoglobin
III. Capable of amoeboid movement

(A) I only

(B) II only

(C) III only

(D) I and III

(E) I, II and III

17. The correct sequence of the phases of mitosis is:

(A) Interphase, prophase, telophase, anaphase

(B) Metaphase, prophase, telophase, anaphase

(C) Telophase, interphase, metaphase, prophase

(D) Prophase, anaphase, interphase, metaphase

(E) Prophase, metaphase, anaphase, telophase

18. The conversion of atmospheric nitrogen (N_2) by certain eubacteria and blue green bacteria into a form usable by other living organisms is termed:

(A) Ammonification

(B) Denitrification

(C) Nitrification

(D) Nitrogen fixation

(E) Transformation

19. Crosses among radishes of three phenotypes produced the following results:

Cross	Parents	Offspring (types of radishes)
1	Oval x Round	156 Oval, 154 Round
2	Long x Round	301 Oval
3	Long x Oval	153 Long, 150 Oval

In a cross between two plants of the long type, what percentage of plants producing long radishes would be expected?

(A) 0

(B) 25

(C) 50

(D) 75

(E) 100

20. CO_2 is released into the atmosphere by which of the following processes?

I. Alcoholic fermentation
II. Photosynthesis
III. Cellular respiration

(A) I only
(B) II only
(C) III only
(D) I and II
(E) I and III

21. The RNA coded for by DNA in the nucleus differs from the mRNA in the cytoplasm in that it contains:

(A) Exon coded segments
(B) Intron coded segments
(C) Uracil
(D) Ribose
(E) Thymine

22. In which cross would 1 / 2 of the offspring be expected to show the dominant phenotype?

(A) Dd x DD
(B) DD x dd
(C) Dd x dd
(D) Dd x Dd
(E) dd x dd

23. Growth of a dicot stem in length results from cells produced in the

(A) apical meristem
(B) cork cambium
(C) endodermis
(D) pericycle
(E) vascular cambium

24. During contraction, the shortening of a skeletal muscle results from the interaction of what two proteins?

(A) Actin and melanin
(B) Actin and keratin
(C) Actin and myosin
(D) Keratin and myosin
(E) Keratin and melanin

25. Which answer below lists the organisms in a sequence that corresponds to the following:

nerve network → nerve ladder → ventral nerve cord with ganglia → dorsal hollow nerve cord?

1. planaria
2. earthworm
3. *Hydra*
4. human

(A) 1–2–3–4
(B) 3–1–2–4
(C) 3–2–1–4
(D) 1–3–2–4
(E) 1–2–4–3

Questions 26 –27 refer to the following sequence.

Algae insect larvae fish humans

26. The sequence of energy flow from one trophic level to another best illustrates:

(A) a food pyramid
(B) succession
(C) a niche
(D) a food chain
(E) a food web

27. In the above sequence the insect larvae would be termed:

(A) Autotrophs
(B) Carnivores
(C) Consumers
(D) Decomposers
(E) Producers

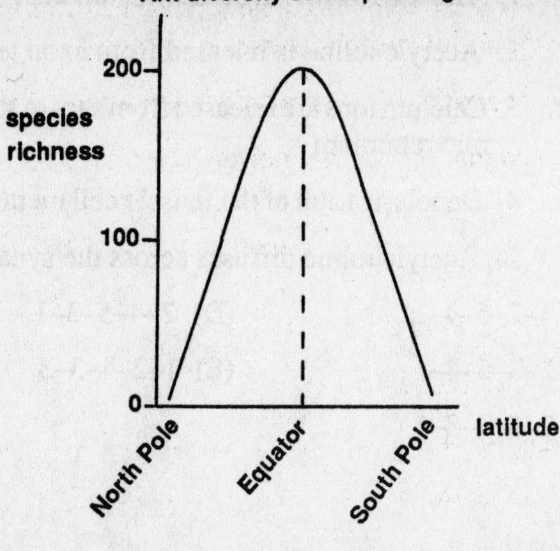

28. Based on the information above, what could you conclude about species diversity on land?

    (A) Species diversity is greatest at latitudes north and south of the equator.

    (B) Species diversity increases at higher latitudes.

    (C) Species diversity decreases as one travels toward the equator.

    (D) Species diversity remains constant as one travels north or south of the equator.

    (E) Species diversity is greatest at latitudes near the equator.

29. In which one of the following biomes would the conclusion in question #28 not apply?

    (A) Taiga

    (B) Tropical desert

    (C) Tundra

    (D) Temperate deciduous forest

    (E) Tropical rain forest

30. The correct sequence of the following events in skeletal muscle contractions is:

1. Thick and thin myofilaments interact

2. Acetylcholine is released from axon terminals

3. Calcium ions are released from sacs of the sarcoplasmic reticulum

4. Depolarization of the muscle cell membrane occurs

5. Acetylcholine diffuses across the synaptic cleft

(A) 3–2–5–4–1

(B) 2–3–5–4–1

(C) 2–5–4–3–1

(D) 2–4–5–3–1

(E) 1–2–4–3–5

31. A deficiency of which vitamin causes rickets in children?

(A) Vitamin A

(B) Vitamin C

(C) Vitamin D

(D) Vitamin E

(E) Vitamin K

32. According to the symbiont theory:

(A) Certain prokaryotic cells gradually changed to form eukaryotic cells.

(B) Eukaryotic cells arose when certain prokaryotic cells entered and "took up residence" inside other prokaryotic cells.

(C) The earliest cells were eukaryotic.

(D) Certain prokaryotic cells gradually degenerated to form viruses.

(E) Certain eukaryotic cells gradually degenerated to form prokaryotic cells.

33. The correct sequence in which food passes through the digestive tract is:

(A) Mouth cavity-stomach-pharynx-esophagus-small intestine.

(B) Mouth cavity-pharynx-stomach-esophagus-small intestine.

(C) Mouth cavity-esophagus-pharynx-stomach-small intestine.

(D) Mouth cavity-pharynx-esophagus-stomach-small intestine.

(E) Mouth cavity-pharynx-esophagus-small intestine-stomach.

34. The hormone that stimulates release of milk and contraction of smooth muscle during childbirth is:

(A) Cortisol
(B) Glucagon
(C) Oxytocin
(D) Prolactin
(E) Testosterone

Questions 35 – 36:

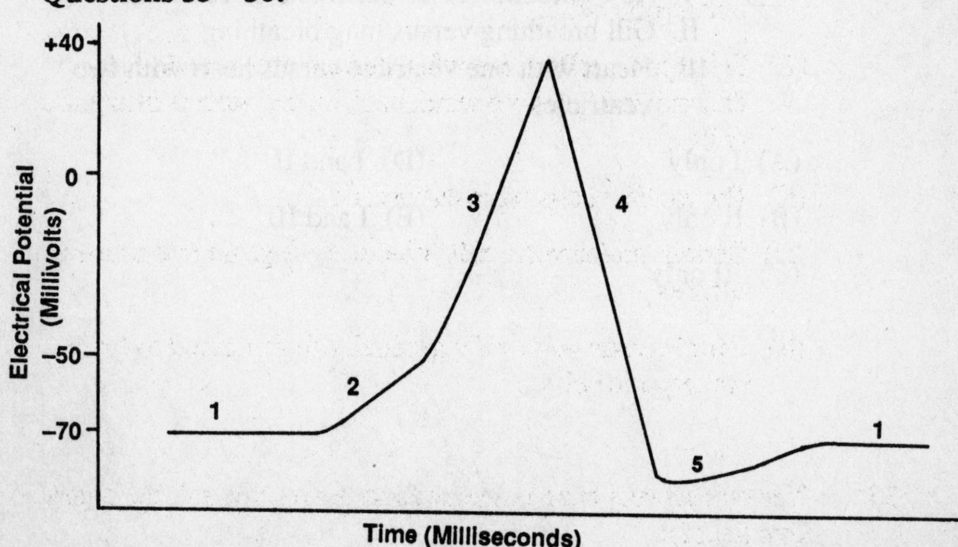

35. The polarity change in an axon membrane potential that occurs at #3 during an action potential can best be explained by a:

(A) Rapid inflow of Cl^-

(B) Rapid inflow of Na^+

(C) Rapid outflow of K^+

(D) Rapid outflow of Ca^{++}

(E) Rapid outflow of Na^+

36. The axon membrane at #4 is:

(A) At resting potential

(B) Depolarized

(C) Hyperpolarized

(D) Repolarizing

(E) Being stimulated at the threshold level

37. A classification scheme that places crocodiles, robins, and foxes in one group and perch and frogs in another group is based on:

I. No backbone versus backbone of vertebrae
II. Gill breathing versus lung breathing
III. Heart with one ventricle versus heart with two ventricles

(A) I only

(B) II only

(C) III only

(D) I and II

(E) I and III

38. Which of the following correctly indicates the species name for a wolf?

(A) *canis lupus*
(B) Canis Lupus
(C) *Canis lupus*
(D) canis lupus
(E) *Canis Lupus*

39. Which of the following develops from embryonic endoderm?

(A) Circulatory system
(B) Lining of digestive system
(C) Muscles
(D) Skin epidermis
(E) Spinal cord

40. Microtubules are associated with the structure of all of the following except the:

(A) Centriole
(B) Cilium
(C) Eukaryotic flagellum
(D) Mitochondrion
(E) Spindle fiber

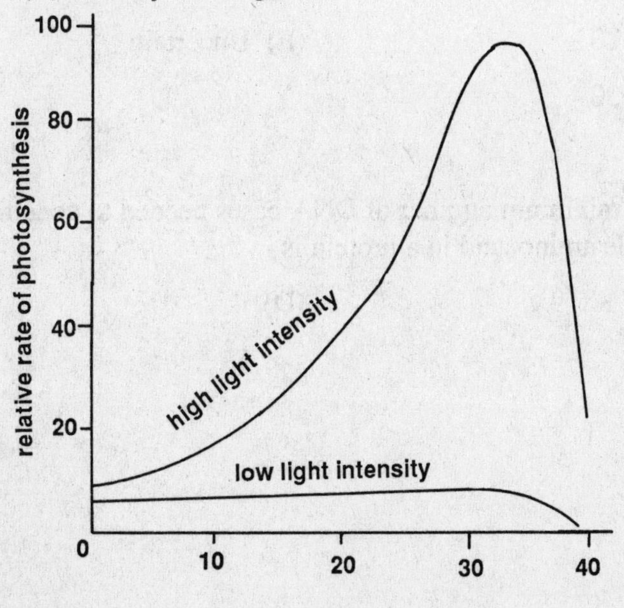

41. Based on data in the figure above, at which of the following light and temperature combinations is the rate of photosynthesis greatest?

(A) Low light intensity; 20°C

(B) High light intensity; 20°C

(B) High light intensity; 30°C

(D) High light intensity; 34°C

(E) High light intensity; 38°C

42. A phylum in the animal kingdom is equivalent to what category in the plant kingdom?

(A) Class

(B) Division

(C) Family

(D) Genus

(E) Order

43. If guanine composes 30 percent of the nitrogen bases in a DNA molecule, what percentage of the bases would be adenine in the molecule of mRNA transcribed from the sense strand?

(A) 10

(B) 20

(C) 30

(D) 40

(E) Uncertain

44. The minimum number of DNA bases needed to code for a single amino acid in a protein is

(A) 1

(D) 4

(B) 2

(C) 3

(E) 5

Questions 45 – 46

The Genetic Code

(Based on Messenger RNA Codons)

First Base	U	C	A	G	Third Base
U	Phenylalanine	Serine	Tyrosine	Cysteine	U
	Phenylalanine	Serine	Tyrosine	Cysteine	C
	Leucine	Serine	Stop	Stop	A
	Leucine	Serine	Stop	Tryptophan	G
C	Leucine	Proline	Histidine	Arginine	U
	Leucine	Proline	Histidine	Arginine	C
	Leucine	Proline	Glutamine	Arginine	A
	Leucine	Proline	Glutamine	Arginine	G
A	Isoleucine	Threonine	Asparagine	Serine	U
	Isoleucine	Threonine	Asparagine	Serine	C
	Isoleucine	Threonine	Lysine	Arginine	A
Start	Methionine	Threonine	Lysine	Arginine	C
G	Valine	Alanine	Aspartic acid	Glycine	U
	Valine	Alanine	Aspartic acid	Glycine	C
	Valine	Alanine	Glutamic acid	Glycine	A
	Valine	Alanine	Glutamic acid	Glycine	G

45. Given the Genetic Code above and the sequence of bases on a sense strand of DNA is 3' CAT TAG CAT TAG 5', what would the two types of amino acids be in the translated amino acid chain?

(A) Isoleucine, leucine

(B) Isoleucine, threonine

(C) Valine, isoleucine

(D) Valine, threonine

(E) Valine, methionine

46. If a mutation results in the substitution of C for G in the DNA triplet 3' ACG 5', then the amino acid _____ would be built into the amino acid chain during translation instead of the amino acid _______.

(A) Leucine, arginine
(B) Tryptophan, arginine
(C) Tryptophan, proline
(D) Tryptophan, cysteine
(E) Arginine, leucine

Questions 47 – 48

ABO Blood Group
Agglutination Results

Individual	Antiserum	
	Anti -A	Anti-B
1	+	–
2	–	–
3	+	–
4	–	+
5	+	+

47. From the results shown above, which individual has blood group O?

(A) 1
(B) 2
(C) 3
(D) 4
(E) 5

48. The individual above with genotype I^AI^B is?

(A) 1
(B) 2

(C) 3

(D) 4

(E) 5

Questions 49 – 50

Gandalf obtained the following results by breeding knobby and flat-kneed hobbits.

	Parents		Offspring	
Cross	Female	Male	Knobby Knees	Flat Knees
1	Knobby	Knobby	98	0
2	Flat	Flat	0	102
3	Knobby	Flat	57	52
4	Knobby	Knobby	76	24
5	Knobby	Flat	101	0

49. In which of the above individuals can you be reasonably certain the genotype is homozygous dominant?

(A) Male of cross 1

(B) Male of cross 2

(C) Female of cross 3

(D) Male of cross 4

(E) Female of cross 5

50. If two knobby kneed hobbits resulting from cross 5 mate with each other, what percentage of their offspring would be expected to have knobby knees?

(A) 100

(B) 75

(C) 50

(D) 25

(E) 0

PART B

Questions 51 – 53:

(A) Commensal

(B) Parasite

(C) Predator

(D) Prey

(E) Scavenger

51. A crow feeds on a dead rabbit lying on the road. The crow is a

52. A barnacle is moved from one feeding place to another while attached to the shell of a horseshoe crab. The barnacle is a

53. A leech feeds on blood after attaching to a swimmer in a freshwater pond. The leech is a

Questions 54 – 56.

(A) Anaphase

(B) Interphase

(C) Metaphase

(D) Prophase

(E) Telophase

54. A characteristic number of double strand chromosomes appear and spindle formation begins

55. A distinct nucleus is evident throughout this phase

56. The nuclear membrane and nucleoli reappear and chromosomes uncoil and spread out as chromatin.

Questions 57 – 59:

(A) Natural active acquired immunity

(B) Natural passive acquired immunity

(C) Artificial active acquired immunity

(D) Artificial passive acquired immunity

(E) Species immunity

57. Type of immunity resulting from vaccinations

58. Type of immunity that results from having an infectious disease

59. Short lived immunity conferred on the newborn as a result of the transfer of antibodies across the placenta

Questions 60 – 62:

(A) Annelida

(B) Arthropoda

(C) Chordata

(D) Cnidaria

(E) Echinodermata

60. Animals in this group have a notochord and pharyngeal gill slits at some stage and form a dorsal hollow nerve cord.

61. Phylum of segmented animals with a complete digestive tube, complete blood circulation and nephridia.

62. Adults in this phylum exhibit radial symmetry and form a water vascular system with tube feet.

Questions 63 – 65:

(A) Grassland

(B) Taiga

(C) Temperature deciduous forest

(D) Tropical rain forest

(E) Tundra

63. Biome characterized by permafrost and the absence of trees.

64. Temperature changes very little throughout the year in this biome which contains a tremendous number and variety of plants.

65. Many trees in this biome shed their leaves in the fall.

Questions 66 – 68:

(A) Chloroplast
(B) Golgi complex
(C) Mitochondrion
(D) Lysosome
(E) Ribosome

66. Flattened sacs called thylakoids occur in stacks called grana inside this organelle.

67. Organelle in which the enzymes of the Krebs Cycle are located.

68. Organelle on which amino acids are linked together to build proteins.

Questions 69 – 71:

(A) Antibiotic
(B) Antibody
(C) Antigen
(D) Complement
(E) Interferon

69. A protein produced by virus infected vertebrate cells which spreads to neighboring cells and helps protect them from the virus.

70. A protein, produced in the body of a vertebrate, that reacts with the foreign protein, or polysaccharide that stimulated its formation.

71. A substance produced by a microorganism which is used to kill or inhibit the growth of other microorganisms.

Questions 72 – 74:

(A) Bacteria
(B) Fungi
(C) Prions
(D) Viroids
(E) Viruses

72. Organisms with prokaryotic cells that contain DNA and RNA.

73. Proposed infectious particles that consist only of protein.

74. Infectious particles composed of naked, circular molecules of RNA.

Questions 75 – 77:

(A) Photosynthetic
(B) Autotrophs
(C) Parasitic
(D) Prokaryotic
(E) Eukaryotic

75. All bacteria.

76. All animals.

77. All fungi.

Questions 78 – 80:

(A) Transcription
(B) Transduction
(C) Transformation
(D) Translation
(E) Transpiration

78. Messenger RNA directs the synthesis of a chain of amino acids.

79. Bacterial genes are transferred from one bacterium to another by a bacteriophage.

80. The sequence of nitrogen bases on a strand of DNA determines the sequence of bases in an RNA molecule.

Questions 81 – 83:

(A) Chloroplast and nucleic acid only
(B) Mitochondrion only
(C) Nucleic acid only
(D) Mitochondrion and nucleic acid only
(E) Mitochondrion, nucleic acid, and chloroplast

81. Found in plant cells.

82. Found in viruses.

83. Found in animal cells.

PART C

Questions 84 –89 refer to the figure below.

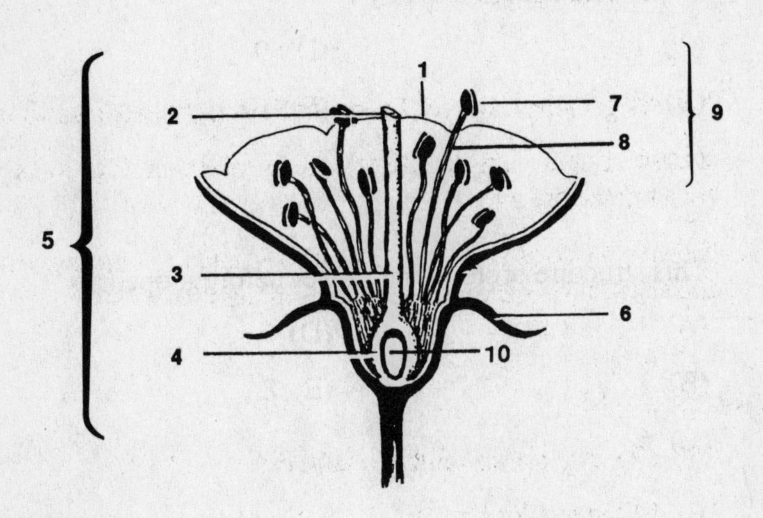

84. Fertilization occurs at

(A) 2
(B) 3
(C) 7
(D) 8
(E) 10

85. Stamens are indicated by

(A) 3
(B) 6
(C) 8
(D) 9
(E) 10

86. Meiosis occurs in

(A) 3 only
(B) 7 only
(C) 10 only
(D) 3 and 7
(E) 7 and 10

87. Pollinators are attracted by

(A) 1
(B) 5
(C) 6
(D) 9
(E) 10

88. This structure protects the flower before it opens

(A) 1
(B) 3
(C) 6
(D) 7
(E) 8

89. Pollen tubes grow through

(A) 1
(B) 3
(C) 6
(D) 7
(E) 8

PART D

Questions 90 – 91:

The arrangement used and the initial contents of five beaker and sac combinations are given below:

Set Up	Beaker Contents	Sac Contents
1	Water	Starch + Amylase + Iodine
2	Starch +Water	Amylase + Iodine
3	Water	Starch + Boiled Amylase + Iodine
4	Amylase + Water	Starch + Iodine
5	Water	Amylase + Iodine

90. After 45 minutes, in which setup would you expect the beaker contents to show a positive test for sugar when heated with Benedict's reagent?

(A) 1

(B) 2

(C) 3

(D) 4

(E) 5

91. In which of the above setups would the sac contents be black in color at the start?

(A) 1 only

(B) 2 only

(C) 4 only

(D) 1 and 4

(E) 1, 3 and 4

Questions 92 – 95:

The effect of various altitudes on the partial pressure of oxygen (P_{O2}), the percentage of blood by volume occupied by corpuscles, and the percent saturation of hemoglobin at various oxygen partial pressures at 38°C are shown in the graph and table.

The effect of altitude on P_{02}

Altitude (ft. above sea level)	P_{O_2} in arterial blood (mm Hg 0	% Red blood Cells
0	100	46. %
2,000	92	48. %
4,000	85	51. %
6,000	79	53. %
8,000	74	56. %
10,000	69	58. %
20,000	35	61. %
30,000	19	

(mmHg)

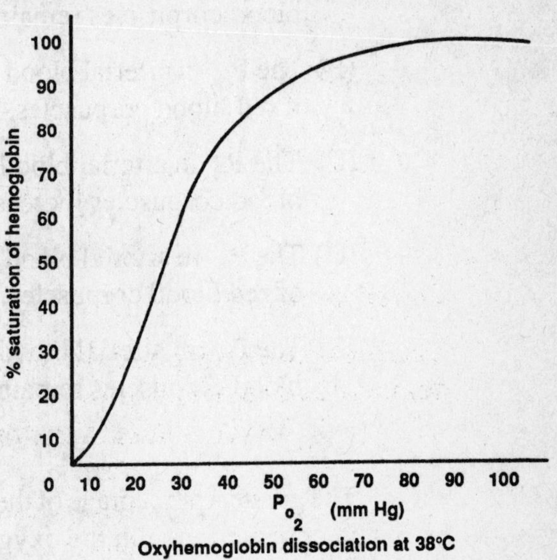

Oxyhemoglobin dissociation at 38°C

92. What would the approximate percent saturation of hemoglobin with O_2 be when the oxygen partial pressure (P_{O_2}) in arterial blood equals 100 mm Hg?

(A) 40

(B) 50

(C) 75

(D) 90

(E) 100

93. Everything else being equal what would the approximate percent saturation of arterial blood hemoglobin with O_2 be at an elevation of 20,000 feet?

(A) 40

(B) 50

(C) 70

(D) 90

(E) 100

94. Which of the following statements would be true at higher elevations?

(A) The P_{O_2} in arterial blood increases but the number of red blood corpuscles remains the same.

(B) The P_{O_2} in arterial blood decreases and so does the number of red blood corpuscles.

(C) The P_{O_2} in arterial blood decreases and the number of red blood corpuscles increases.

(D) The P_{O_2} in arterial blood increases and so does the number of red blood corpuscles.

(E) The P_{O_2} in arterial blood decreases and the number of red blood corpuscles remains the same.

95. The greatest percentage of the oxygen transport in blood occurs in conjunction with the oxygen transporting pigment, hemo-

globin in the:

(A) Plasma

(B) Platelets

(C) Red blood cells

(D) Thrombocytes

(E) White blood cells

Questions 96 –97:

Using data on the average height of males and females from birth to age 18 shown in the graph below, answer the next two questions.

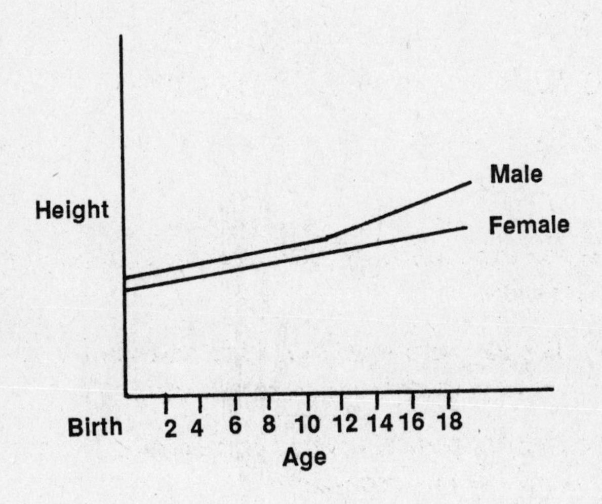

96. The average height data in the above graph indicates that:

(A) Females are shorter than males at all ages.

(B) Males increase in height faster than females between ages 10 and 12.

(C) Most females are shorter than 160 cm.

(D) Males continue to grow in height longer than females.

(E) Females increase in height faster than males between ages 14 and 16.

97. Beyond age 6 the greatest increase in average height of males occurs between what age groups?

(A) 6 and 8
(B) 8 and 10
(C) 10 and 12
(D) 12 and 14
(E) 14 and 16

Questions 98–100 are based on the figure and the data on the relative concentration of selected substances from three locations (A, B, and C) in the kidneys.

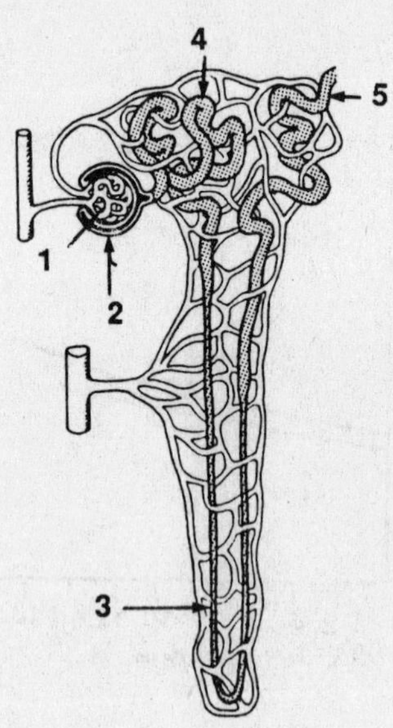

98. The composition of fluids obtained at Locations A and B was analyzed and found to be similar. Locations A and B in Table 1 correspond to what two locations illustrated in the Figure?

(A) 1 and 2
(B) 1 and 3
(C) 1 and 5
(D) 2 and 3
(E) 2 and 5

99. The absence of glucose at Location C is most likely a result of:

(A) Excretion
(B) Filtration
(D) Reabsorption
(E) Secretion

(C) Osmosis

100. The concentration of substances found at Location C in Table 1 would most likely occur at what location in the figure?

(A) 1

(B) 2

(C) 3

(D) 4

(E) 5

BIOLOGY
TEST VI

ANSWER KEY

1.	A	34.	C	67.	C
2.	D	35.	B	68.	E
3.	E	36.	D	69.	E
4.	D	37.	C	70.	B
5.	E	38.	C	71.	A
6.	B	39.	B	72.	A
7.	E	40.	D	73.	C
8.	C	41.	D	74.	D
9.	A	42.	B	75.	D
10.	E	43.	E	76.	E
11.	E	44.	C	77.	D
12.	C	45.	C	78.	D
13.	C	46.	D	79.	B
14.	C	47.	B	80.	A
15.	C	48.	E	81.	E
16.	D	49.	E	82.	C
17.	E	50.	B	83.	D
18.	D	51.	E	84.	E
19.	E	52.	A	85.	D
20.	E	53.	B	86.	E
21.	B	54.	D	87.	A
22.	C	55.	B	88.	C
23.	A	56.	E	89.	B
24.	C	57.	C	90.	A
25.	B	58.	A	91.	E
26.	D	59.	B	92.	E
27.	C	60.	C	93.	C
28.	E	61.	A	94.	C
29.	B	62.	E	95	C
30.	C	63.	E	96.	D
31.	C	64.	D	97.	D
32.	B	65.	C	98.	A
33.	D	66.	A	99.	D
				100.	E

BIOLOGY
TEST VI

DETAILED EXPLANATIONS OF ANSWERS

1. (A)

DNA and RNA differ in that DNA nucleotides contain deoxyribose and RNA nucleotides contain ribose. In addition, DNA contains thymine and RNA contains uracil. The nitrogen base complement for both thymine and uracil is adenine.

2. (D)

A colorblind male (X^cY) and a carrier female (X^CX) can produce daughters of two genotypes: X^CX^c and X^cX^c.

3. (E)

A taxis is a response to a stimulus of some sort. Some organisms respond to light (phototaxis) or chemicals (chemotaxis).

4. (D)

In a second exposure to an antigen the concentration (titer) of antibody produced is greater and production time is shorter than in a first exposure.

5. (E)

While there are DNA and RNA viruses any given type of virus contains only one type of nucleic acid.

6. (B)

Changes in form are known as metamorphosis. For example, an insect egg may hatch into a caterpillar. The caterpillar spins a cocoon and hatches out as a butterfly. This is complete metamorphosis. Grasshoppers also undergo metamorphosis, but the young resemble the adult form.

Hemaphroditism refers to a condition whereby an organism such as an earthworm has both male and female structures and is fertile. Parthenogenesis is development of an egg without fertilization. Transduction relates to the incorporation of bacterial DNA into bacteriophage DNA and the subsequent transfer of the bacterial DNA to other bacteria. Transformation relates to the changing of a plant cell's genetic makeup using bacteria.

7. (E)

8. (C)

With the exception of an occasional mutation all cells produced by mitosis are genetically identical. In contrast, all cells produced by meiosis are genetically different and contain a reduced (monoploid or haploid) chromosome number.

9. (A)

Carbon dioxide is incorporated into organic compounds during the Calvin cycle (dark reactions of photosynthesis) when it combines with ribulose diphosphate.

10. (E)

When populations of a single ancestral type change over time to form several distinct species both divergent evolution and adaptive radiation have occured.

11. (E)

A yeast cell and an elephant cell are both eukaryotic. Neither contains chlorophyll. Thus, they must both be heterotrophic.

12. (C)

Nitrogen occurs in all amino acids, which are the basic building blocks of proteins, but does not occur in all carbohydrates and lipids. All 3 biomolecules contain carbon, hydrogen, and oxygen. Phospholipids contain phosphorus.

13. (C)

While free swimming sperm cells are characteristic of ferns, in flowering plants a generative nucleus inside a pollen grain divides to form two nonmotile sperm which are brought close to the egg by pollination and growth of a pollen tube.

14. (C)

Citric acid is formed when the acetyl group and oxaloacetate combine. Each of the acids listed is a component in the Krebs Cycle.

15. (C)

In the inheritance of a two allele system the frequency of the two alleles (dominant versus recessive) is represented by the equation $p + q = 1$ and the frequencies of the genotypes (homozygous dominant versus heterozygous versus homozygous recessive) is represented by $P^2 + 2pq + q^2 = 1$. Since 1,600 out of 10,000 individuals sampled showed the recessive phenotype and are homozygous recessive, the frequency of the recessive genotype, q^2, is determined as follows: $q^2 = 1{,}600 / 10{,}000 = .16$ (or 16%). The frequency of the recessive allele, q, is determined by taking the square root of q^2:

$q = \sqrt{q^2} = \sqrt{.16} = .4$ The frequency of the dominant allele, p, is determined by subtracting q from 1: $p = 1 - q = 1 - .4 = .6$.

Finally, the frequency of homozygous dominant individuals, p^2, is determined by squaring P: $p^2 = .6^2 = .36$ (or 36%).

16. (D)

Mature white blood cells or leukocytes are nucleate and exhibit ameboid movement.

17. (E)

Mitosis, the division of the nucleus, is subdivided into prophase, metaphase, anaphase, and telophase. Interphase is the stage during which the cell growth and synthesis occur.

18. (D)

The conversion of atmospheric nitrogen into substances generally usable by living organisms is termed nitrogen fixation.

19. (E)

The crosses illustrated involve intermediate inheritance or incomplete dominance. Both the long, LL, and round, L^1L^1, phenotypes are homozygous. The heterozygous condition, LL^1, results in the oval phenotype. A cross between two long plants (LL x LL) would be expected to produce only long radishes.

20. (E)

In both alcoholic fermentation and cellular respiration CO_2 is formed while in photosynthesis it is incorporated into organic compounds.

21. (B)

In eukaryotes, the precursor RNA transcribed from DNA in the nucleus contains both exon and intron coded segments. Nucleotide sequences in introns are not translated to protein. Enzymes remove the intron coded segments from the precursor RNA and splice the exon coded segments back together to form mRNA.

22. (C)

Using a Punett square illustrates this best.

	D	d
d	Dd	dd
d	Dd	dd

23. (A)

Cells of the apical meristein divide and subsequently elongate, thus increasing stem length. The cork cambium and vascular cambium result in growth in width or compensate for such growth. The pericycle is that group of cells which forms lateral roots. The endodermis gives the plant selective uptake of minerals from the soil.

24. (C)

According to the sliding filament hypothesis of skeletal muscle contraction, thin filaments of actin slide along thick filaments of myosin from both ends of the thick filaments toward the middle of each sarcomere.

25. (B)

26. (D)

The series of organisms (producers, consumers, etc.) through which energy and nutrients move in a community is called a food chain. The relationships are often more complex since any given organism may feed on or be fed on by several other kinds of organisms. A food web is used to illustrate the network of interrelationships between organisms.

27. (C)

In general, any organism that eats another organism is called a consumer. Since the insect larvae feed on producers they could also be termed herbivores.

28. (E)

Species diversity is greatest in tropical rainforests located on or near the equator.

29. (B)

Due to the lack of adequate precipitation in desert regions, the vegetation is characteristically scarce as is the animal life.

30. (C)

31. (C)

Some of the problems caused by deficiency of the other vitamins includes nightblindness (deficient vitamin A), improper fatty acid metabolism (deficient vitamin E), delayed blood clotting (deficient vitamin K), and scurvy (deficient vitamin C).

32. (B)

Many biologists believe that eukaryotic cells evolved when formerly free living prokaryotes established a symbiotic relationship. Perhaps one prokaryotic cell ingested another and instead of digesting it began to coexist with it. Much of the evidence for the symbiant theory involves similarities between bacteria and two of the main organelles of eukaryotic cells: chloroplasts and mitochondria.

33. (D)

34. (C)

Oxytocin, which is produced by the hypothalamus and released from the posterior pituitary gland, stimulates milk ejection from the breasts and uterine muscle contraction both during and after childbirth.

35. (B)

During an action potential in a nerve fiber, depolarization, which is depicted at #3, occurs as a result of a rapid inflow of sodium ions (Na^+).

36. (D)

At #4 the axon membrane is repolarizing as a result of a rapid outflow of potassium ions (K^+). Resting potential is indicated by 1. Number 2 represents the threshold level and number 5 represents hyperpolarization. Depolarization occurs between 4 and 5.

37. (C)

All of the animals listed are vertebrates with a backbone consisting of vertebrae. Only the perch uses gills to breath throughout its life. A frog begins life as a gill breathing tadpole which metamorphoses into a lung breathing adult. Finally, the heart of a perch consists of one atrium and one ventricle, the frog heart of two atria and one ventricle, and, the four chambered heart of crocodiles, robins, and foxes consists of two atria and two ventricles.

38. (C)
The species name has two parts. The first part of the name, i.e. *Canis*, represents the genus in which the species is classified and is capitalized. The second part of the name, *lupus*, is the specific epithet and is lowercase. Both parts are italicized or underlined in print.

39. (B)
The ectoderm leads to formation of the epidermis and the entire nervous sytem. Embryonic mesoderm forms muscles and components of the circulatory system.

40. (D)
Both a cilium and a eukaryotic flagellum contain nine double microtubules arranged in a circle near the periphery and two separate microtubules in the center of the circle. . .a "9 + 2" arrangement. A centriole contains nine triplets of microtubules arranged in a circle near the periphery. Each spindle fiber is one microtubule.

41. (D)

42. (B)

43. (E)
Unless the base composition of the sense strand of DNA that codes for the mRNA is known the answer cannot be determined with certainty. In addition, intron coded RNA segments are not incorporated into mRNA.

44. (C)

Coding for amino acids is based on a triplet code. The "alphabet" contains 4 letters (bases) and the words contain 3 letters (bases) each. Thus you can have 64 different codes.

45. (C)

The sequences of bases on the segment of mRNA transcribed would be 5' GUA AUC GUA AUC 3'. The codon "GUA" codes for the amino acid valine and the codon "AUC" codes for isoleucine.

46. (D)

The DNA triplet "ACG" is transcribed to the mRNA codon "UGC" which is, in turn, translated to the amino acid cysteine. If a mutation results in the substitution of C for G in the DNA triplet, the new DNA triplet "ACC" is transcribed to the mRNA codon "UGG" which is translated to the amino acid tryptophan.

47. (B)

Since an individual with type O has neither antigen A or B in their red blood cells, agglutination will not occur in the presence of either Anti-A or Anti-B.

48. (E)

An individual with the genotype $I^A I^B$ forms both A and B antigens. So, agglutination would occur in the presence of both Anti-A and Anti-B.

49. (E)

The male in cross 1 is probably homozygous dominant. However, the female in cross 5 is no doubt homozygous dominant. In this case, a test cross has been made. A parent with the dominant phenotype was crossed with the recessive (homozygous, of course). In such crosses, if all the offspring are of the dominant phenotype, the parent is homozygous dominant. If about 1/2 the offspring display the recessive trait, the parent is heterozygous recessive. Cross 3 would be an example of a heterozygous parent.

50. (B)

The offspring of Cross 5 are all heterozygous knobby (Kk). In a cross between two heterozygous knobby knee individuals (Kk x Kk), the probability of producing offspring with knobby knees is 3/4 or 75%.

51. (E)

A scavenger is an animal that feeds on dead animals.

52. (A)

A commensal is an organism that benefits by existing in a symbiotic relationship with another type of organism without harming or benefitting it.

53. (B)

A parasite is an organism that gains its existence by feeding on and, to some degree, harming another type of living organism, its host. A predator seeks and kills another animal for food. The animal which is hunted by a predator is called the prey.

54. (D) 55. (B) 56. (E)

The various phases of mitosis have characteristics which allow us to identify them. During metaphase the chromosomes are lined-up along the metaphase plate or equator of the cell. In anaphase, the chromosomes are pulled away from the equator of the cell.

57. (C)

A vaccine is a preparation containing one or more antigens which stimulates an immune response (i.e. antibody production, etc.) following vaccination. The immunity is typically long lasting.

58. (A)

An antigen (i.e. component of a microorganism) encountered by the body during the course of an infectious disease stimulates an immune response (i.e. antibody production, etc.). The immunity is typically long lasting.

59. (B)

Certain antibodies can cross the placenta late in pregnancy to confer on the newborn resistance to diseases to which the mother is immune. Since the newborn cannot form the antibodies, the immunity is short lived.

60. (C) 61. (A) 62. (E)

The phylum anthropoda includes insects. These organisms have chitinous exoskeletons and jointed appendages. The phylum cnidaria includes organisms such as hydra.

63. (E) 64. (D) 65. (C)

Grasslands are characterized by short, herbaceous, (grassy) plants. The taiga has a relatively long harsh winter, cool summer and an abundance of pine or evergreen trees.

66. (A) 67. (C) 68. (E)

Chloroplasts consist of grana and stroma. Each granum is a stack of flattened sacs called thylakoids. The mitochondrion is the powerhouse of the cell. Lysosomes contain digestive enzymes. The Golgi apparatus plays a role in packaging of nutrients within the cell. Ribosomes free in the cytoplasm or lined up along side the endoplasmic reticulum serve as sites of protein synthesis.

69. (E) 70. (B) 71. (A)

Interferons are proteins produced by virus infected cells which spread to neighboring cells to cause inhibition of virus multiplication.

An antibody (immunoglobulin) is a glycoprotein produced in the body of a vertebrate in response to an antigen (i.e. a component of a microorganism, a pollen grain, a fungal spore, food, etc.). The antibody usually reacts with the antigen. For example, antitoxins are antibodies that react with toxins.

Natural, semisynthetic, and synthetic antibiotics are used today as drugs mainly in treating infectious diseases caused by bacteria. Generally, they are effective in killing or inhibiting microorganisms at very low concentrations.

72. (A) 73. (C) 74. (D)

The cells of both eukaryotes and prokaryotes contain both DNA and RNA.

Although his work is somewhat controversial, Dr. Stanley Pruisner proposed that several diseases including scrapie in sheep and goats and Creutzfeldt -Jacob disease in humans are caused by protein infectious agents or prions.

Viroids, which are known to cause certain plant diseases, consist of naked, circular molecules of RNA.

75. (D) 76. (E) 77. (D)

All bacteria have prokaryotic cells. Some bacteria are autotrophic (photosynthetic, chemosynthetic). Autotrophs can make organic molecules from simple inorganic molecules and an outside energy source. Others are heterotrophic (parasitic, saprophytic, etc.). Heterotrophs require an organic carbon source for energy.

All animals are heterotrophs with eukaryotic cells.

All fungi are heterotrophs with eukaryotic cells.

78. (D) 79. (B) 80. (A)

When the sequence of bases on mRNA codes for the sequence of amino acids in a polypeptide the process is called translation.

On occasion, in a lytic cycle, a bacteriophage (virus that infects bacteria) may be produced inside the host bacterium which incorporates a few bacterial genes in addition to its own bacteriophage genes. When the bacteriophage attaches and injects its DNA into another host bacterium, it also introduces the bacterial genes.

When the sequence of bases on a strand of DNA codes for the sequence of bases in an RNA the process is called transcription.

81. (E) 82. (C) 83. (D)

Plant cells have both mitochondria and chloroplasts. Mitochondria are necessary to carry on cell respiration.

Any given virus is composed of a single type of nucleic acid (DNA or RNA) surrounded by a protein coat or capsid. In some of the more complex viruses an envelope containing lipids and glycoproteins is acquired before or during exiting of the host cell.

84. (E) 85. (D) 86. (E) 87. (A) 88. (C) 89. (B)

Number 10 is an ovule which contains an egg. The pollen tube grows down the style (3) to the ovary (4). The pollen tube penetrates the ovule and releases two sperm nuclei. One fertilizes the egg, the other fertilizes the polar nucleus.

The stamen is indicated by (9) and is composed of an anther (7) and a filament (8).

In flowering plants, meiosis occurs during the formation of pollen inside an anther and in the formation of the egg in the ovule.

The petal (1) is usually colored if pollinators are used to transfer pollen from the anther to the stigma (2). Structure (6) is usually green or dull in color and serves to protect the developing flower before it opens.

Structure (5) is the entire female structure: the pistil or carpel.

90. (A)

Neither starch, a polysaccharide or amylase, a protein can pass through the sac membrane. However, small sugar molecules can. Digestion of starch to simple sugars that pass through the sac membrane and react with Benedict's reagent occurs in Set Up #1.

91. (E)

Starch turns black in color in the presence of iodine.

92. (E)

At a P_{O_2} of 100 mm Hg hemoglobin is approximately 100 percent saturated. As the P_{O_2} decreases the percent saturation of hemoglobin decreases.

93. (C)

At an elevation of 20,000 feet the P_{O_2} in arterial blood is 35 mm Hg and hemoglobin is approximately 70 percent saturated.

94. (C)

As altitude increases the P_{O_2} in arterial blood decreases and the percentage of blood by volume occupied by red blood cells (corpuscles) increases.

95. (C)

The bulk of the oxygen transport (about 97 percent) occurs in conjunction with hemoglobin located inside red blood cells.

96. (D)

97. (D)

Based on data in the table, the greatest increase in average height of males of 13 cm occurs between ages 12 and 14. The next greatest increase of 11 cm occurs between ages 8 and 10.

98. (A)
The glomerular filtrate inside Bowman's capsule (#2) is essentially the same in composition as blood plasma in the afferent arteriole (#1).

99. (D)
Glucose is reabsorbed. That is, it is returned to the blood from the kidney tubules.

100. (E)
The concentration of substances in the distal end of a collecting duct (#5) is essentially the same as in urine.